Your
Body
and
Radiation

CONTENTS

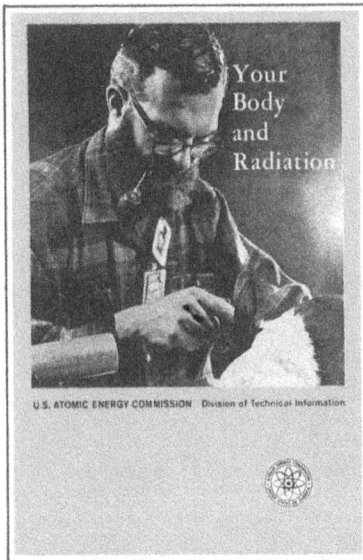

Your
Body
and
Radiation

U.S. ATOMIC ENERGY COMMISSION   Division of Technical Information

NORMAN A. FRIGERIO

THE COVER

and

THE AUTHOR

A live radiobiologist, who also is the author of this booklet, and a rabbit, which is a "phantom", although very much alive, too, are pictured on the cover. Dr. Norman A. Frigerio, of the Argonne National Laboratory, was using the rabbit as a test animal, or "phantom", to determine how a rabbit's body absorbs and reflects neutrons and photons originating in the JANUS Biological Reactor Facility at the Laboratory (see Figure 10 on page 31). The level of the radiation emerging from the instrument at left was sufficient to give the desired information, but too small to affect either the rabbit or the scientist because the dosimeters used for the study (on table and clipped to Dr. Frigerio's shirt) are far more sensitive detectors than either man or rabbit is. Dr. Frigerio has conducted research on the effects of radiation on living organisms at Argonne National Laboratory since 1957. He received his B.S. degree from the Massachusetts Institute of Technology and his Ph. D. in biochemistry from Yale University. His publications, which include books, patents and scientific articles, now number 90; most of them deal with the somatic, or physiological, effects of radiation, in other words: "Your Body and Radiation".

Your Body and Radiation

By NORMAN A. FRIGERIO

1 GROUND RULES

1.10   Since "the proper study of mankind is man"* we will try to steer our tale of radiation effects in that direction. Unfortunately (or fortunately) man makes a rather touchy experimental animal. So, we'll be forced to take most of our information from experience with cells, large and small cooperative mammals, occasional bacteria and other plants, and whatever human incidents or accidents we can find. Also we'll avoid what happens to man's offspring (genetic effects)† and concentrate on radiation's *somatic* or bodily *effects* We'll try to bring in not only the general scientific side (radiobiology) but the beneficial side (radiology) as well.

The art of using a little jargon, or complicated words to express simple ideas is invaluable in any scientific discussion. Not only does it help to impress one's friends, but it actually helps keep track of peculiarly scientific ideas in a forest of conversational ones. "Sodium chloride" conjures up in a chemist's mind images of ions, cubic crystals, and similar useful things. "Salt" is more likely to raise a picture of flat soup.

---

*Alexander Pope—a man himself and so, presumably, qualified to judge—said this in 1734.

†Leaving that for another booklet in this series called *Genetic Effects of Radiation*.

In any case we'll try to keep the jargon to a useful minimum, italicizing new words the first time they're used as we go along. To help follow the game we've added a score card in the form of a Glossary Index that tells where a word is first defined and where else it's been used in the text. Just the same a copy of *Nuclear Terms: A Brief Glossary* from this series will help so much you'll feel you've been cheating. An ordinary unabridged dictionary may be quite a help, too.

Now radiobiology is not as old or as well-scrubbed a field as, say, physics. As a result, there are quite a few more hypotheses* than laws† floating around in it. To keep you well rounded we've chosen to treat all these as contributing something to an understanding of the effects of radiation on people. But, all in all, we've chosen to emphasize what might be called the DNA-LET theory. It's the most popular at the moment and leaves you fairly well prepared to catch up with whatever discoveries tomorrow may bring.

To provide a little better understanding throughout we've cross-referenced ideas back to the section where they're first discussed. For example, "4.31" means look at Section 31 of Chapter 4.

Some of you who are not strongly inclined to physics may find Chapter 2 heavy going. In that case, you might consider starting with Chapter 3, and just come back to Chapter 2 as need arises for clarification of terms, or when the cross-references suggest it.

Finally we've added a list of references for you who like to dig a little deeper.

With that we can be off.

---

*Guesses.
†Guesses that haven't fallen through yet.

2 HOT STUFF, OR A LITTLE ABOUT RADIATION CHEMISTRY AND PHYSICS

2.1 Radiation Types

2.10 Radiation, unlike Caesar's Gaul, can be divided into only two parts, ionizing and exciting. Ionizing radiation is the sort that is of chief interest to us here, no matter how "exciting" the other may be in other fields. This is simply because the *ionization* (or the splitting into charged fragments) of any of the body's atoms, which were not meant to be ionized, creates havoc in the body's well-ordered economy.

Radiation, like good things, comes in small packages — usually called *particles*. Although calling them a quivering cloud might give a better picture for some purposes, we can think of particles here as simply very tiny balls. These too come in two sorts, charged and neutral. The charged particles carry an electrical charge, either positive (+) (*protons, alpha particles, positrons*) or negative (–) (*electrons*). They will be attracted to other charged particles (those that make up the matter they approach) of the opposite sign (+– or –+) or repelled by those of the same sign (++ or ––). This electrical force is so strong that charged particles penetrate matter (including tissue) very poorly. Their energy is spent so quickly pushing and pulling the protons and electrons they meet near the surface of what they strike that it's all used up before they get far. They act much like strong ball magnets would if rolled into a handful of steel marbles — yanking the negative electrons away from their peaceful (neutral) molecular homes, and leaving the molecules behind in an ionized (charged) condition. Air molecules ionized by energetic particles can be seen in Figure 1.

2.11 The neutral particles, chiefly *neutrons* and *photons,** have no charge to use in this way. Their ability to ionize comes about by direct collision. Much like cue balls on a pool table, they either knock another "pool ball" hard

------

*A photon is often called a *quantum* (a little package) of radiation. This is done to indicate that it, unlike the others, behaves more like a cloud of energy than like a ball of matter.

enough to ionize it by sheer impact, or pretty much miss it completely. Since the particles in molecules are relatively about as far apart as stars in the universe, neutral "cue balls" miss lots more often than they hit. So, they get much farther than their charged brothers before losing all their energy.

Figure 1 *Ionized air molecules produced by a beam of energetic particles emerging from the 60-inch cyclotron at the Argonne National Laboratory. The ionization produces a glow in the air much like the glow of ionized gases in a "neon" sign.*

2.12 Before their true nature was known, beams of helium nuclei (2 protons and 2 neutrons fellow-traveling together) were named *alpha rays*. Beams of electrons or positrons were called *beta rays*, and beams of high-energy photons from radioactively decaying elements were named *gamma rays*. Beams of mixed lower energy photons obtained by slamming electrons into heavy metal targets were called *X rays*. High-energy neutron and proton beams were first produced many years later and never received any special names.

2.13 Each of these particles has its own favorite targets. Charged particles mostly pick on the equally charged electrons orbiting outside the atomic nuclei. This is largely because the electrons occupy and enclose so much more

4

space than the nucleus, much like the planets and asteroids that orbit the sun. Photons also do most of their dirty work on the lighter electrons, but this is largely because the photons are themselves so "light".* A photon hitting a heavy nucleus is rather like a bee hitting a bowling ball. The bee bounces, not the ball!†

Neutrons have about the same mass as hydrogen atoms, so that hydrogen nuclei (protons) are their favorite target. Thus living tissue, rich as it is in H_2O, gets badly punished by neutrons. Other kinds of nuclei get some punishment, but the heavier they are the less they get — the bee and the ball again. Since neutrons are uncharged they tend to brush through electron clouds with some disturbance but with little permanent damage, much like baseballs through a thick fog.

2.14   As these particles lose energy they come to different ends. Electrons and protons simply slow down by progressive bounces until they can undergo chemical reaction with the molecules around them. A proton, for example, eventually becomes either an atom of hydrogen, H, or the common hydrogen ion‡ of water, $H_3O^{\oplus}$.

A positron slows down about as much as an electron does, but instead of simply coming to rest, it ends up by being captured by an electron. The pair then "commits suicide", becoming two new photons, in a process called *annihilation*. Photons themselves are eventually absorbed completely and usually produce heat.

Neutrons are eventually captured by nuclei to produce new and heavier nuclei that are always unstable. These unstable nuclei are said to be *radioactive* (2.8) and they spit out photons and/or charged particles until new but stable nuclei finally result.

2.15   Particles of radiation, like bullets, have little effect unless in rapid motion, that is, unless they possess *kinetic energy*. For reasons of physical convenience particle energies are usually given in units of *electron volts*

---

*Having no weight at all.

†Not invariably. High-energy photons have enough "sting" at times to knock positrons or other particles out of a nucleus.

‡More precisely the hydronium ion. The simple hydrogen ion, $H^{\oplus}$, reacts immediately with H_2O to form $H_3O^{\oplus}$!

(ev), the energy an electron gains in moving from the negative to the positive pole of a 1-volt battery. Units of *kilo-electron volts* (1 kev = 1000 ev) and *mega-electron volts* (1 Mev = 1,000,000 ev) are also used for more energetic particles.* Now even 1 Mev is not a great deal of energy by ordinary standards. Some 26,200,000,000,000 (2.62 $\times 10^{13}$) Mev would be needed to raise the temperature of 1 gram of water 1°C. However, these particles occur in very large numbers. One gram of hydrogen contains over 6 $\times 10^{23}$ electrons, for example. The energy per particle may be small, but with such large numbers of particles, the total energy may be quite large. Then, too, chemical molecules are held together by energies of only a few ev so that the collision of even a 1-kev-particle can cause considerable chemical havoc.†

2.2 Chemical Effects

2.20   Molecules, especially biological ones, are most often held together by electron pair (*covalent*) bonds. In Figure 2 we see the most common of biological molecules, water, being split by ionizing radiation. If the bond is split on one side (A) or the other (B), two ions (charged chemical fragments) result. Such ions are usually quite reactive chemically. Now the splitting of the molecule is, itself, disturbing to the biochemistry of the organism. But the further reactions of the ions damage other molecules nearby as well, adding a sort of insult to injury!

2.21   The bond can also be split "down the middle" (C). In this case the products are not charged and, therefore, not ions. Instead they are called *free radicals*, because they contain at least one unpaired electron. They resemble the $CH_3\cdot$, $C_6H_5\cdot$, etc., "radicals" of familiar chemical formulas. Because of the tendency of these "free" electrons to pair themselves and form covalent bonds, free radicals are much more reactive than ions and the net unpleasantness in tissue is even greater. It is believed that

---

*1 ev = 1.6×10^{-12} erg = 1.6×10^{-14} joule = 3.82×10^{-20} calorie. The complete conversion of one hydrogen atom to energy, according to Einstein's E = mc^2, yields 942 Mev.

†32.5 ev will cause the ionization of almost any molecule in living tissue.

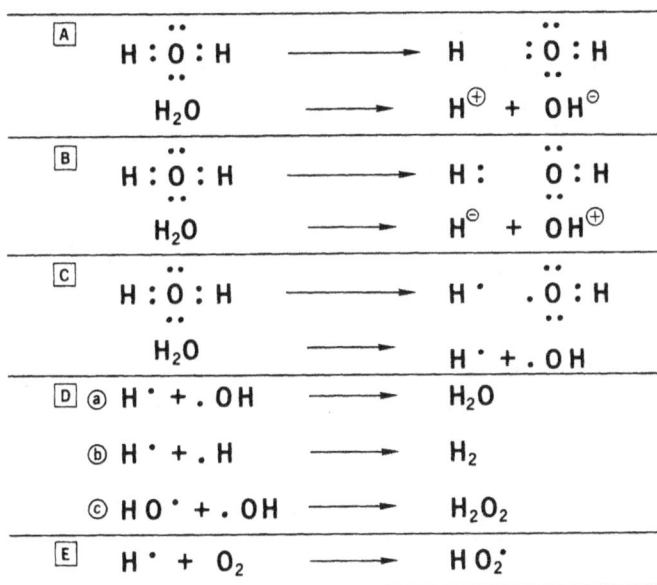

Figure 2 *Chemical effects of ionizing radiation.*

most of the biochemical damage of radiation comes via these free radicals.

Once these radicals are formed there may be enough of them around to react with one another (D). Reforming the original molecule (Da) leaves everything pretty much as before. Formation of hydrogen (Db), while not pleasant, results in little net damage; a few loose H_2 molecules can be tolerated by the body. However (Dc) leads to formation of hydrogen peroxide, H_2O_2, which is distinctly unhealthy. In fact chemical poisoning by hydrogen peroxide resembles radiation illness in many respects.

At least as bad is the combination of a free radical with oxygen (E). The resulting $HO_2 \cdot$ radical seems to have even more undesirable habits than hydrogen peroxide. In fact experiments have shown that a cell with its normal content of oxygen is about three times more sensitive to gamma rays than one temporarily deprived of its oxygen, probably because of the formation of $HO_2 \cdot$. The oxygen-free cell can form little $HO_2 \cdot$ so that damage from this source is minimized.

2.3 Direct and Indirect Effects

2.30   From this one can see the possibility of two kinds of molecular damage. In the first, called the *direct effect*, a biologically important molecule is struck directly by an incoming particle and split into biologically useless fragments. Probably the most important molecules in the living cell are the *DNA* (deoxyribonucleic acid) molecules of the *cell nucleus*. These carry the master blueprints needed by the cell to reproduce itself properly.* Direct destruction of a DNA molecule results in a cell that can live but not divide. On dying of "old age" it leaves no daughters behind to carry on its work. Such progressive cell death without replacement soon leads to the malfunction and eventual death of the irradiated tissue. If this dying tissue is essential to the organism, and cannot be replaced in time, the entire organism will degenerate and die prematurely.

2.31   An *indirect effect* occurs if a less critical molecule, usually water, is split into reactive ions or radicals. If these reactive fragments then drift over to react with such critical molecules as the DNA of the nucleus, damage will be much the same as if they had been struck directly. In direct action no time lag exists between collision and destruction since the particles themselves travel at nearly the speed of light. With indirect action the diffusion of ions and radicals may be sufficiently slow that chemical protective agents may be placed in their path, and thus sacrificed to protect the most critical molecules. The situation is somewhat analogous to a chess game where pawns, bishops, etc., may be sacrificed to an opponent's equally slow-moving men to protect the king (indirect action). However, even a solid line of pawns would provide little protection if a baseball were thrown directly at the king (direct action)!

2.4 Linear Energy Transfer

2.40   As a charged particle speeds along through tissue it collides with parts of atoms every once in a while, much

---

*This process is described in more detail in *Radioisotopes and Life Processes,* a companion booklet in this series.

like a bullet shot into a thin forest collides with leaves and branches. At each collision about 100 ev of energy is lost and about 10 free radicals and ions are left behind in a little clump called a *spur*.

2.41 Electrons and positrons, being rather light, "ricochet" fairly easily, and their spur tracks are a bit erratic. The spur track of a 1-Mev electron looks like this:

This is the case whether these particles impinge on the tissue as primary "beta rays" or have been secondarily produced by photons (from a primary beam of X or gamma rays) colliding with electrons, as in Figure 3A, 1, and 2, on pages 10 and 11.

Protons and alpha particles, being much heavier, ricochet less easily. They resemble cannon balls more than bullets, and leave short, dense spur tracks. The track of a 1-Mev proton looks like this:

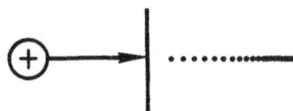

This is the case, again, whether a primary proton is involved or a secondary one produced by a primary fast neutron as in Figure 3B, 1, and 2.

2.42 Two features of this behavior are significant: (a) For a given energy loss (1 Mev in this case), an electron travels much farther than a proton so that the spacing between electron spurs is greater; and (b) as either particle reaches the end of its track *spur spacing* decreases.

In general, spur spacing decreases with increasing particle mass, increasing particle charge, and decreasing particle energy.

2.43 In radiobiology, particles can be compared on the basis of their average *Linear Energy Transfer* (*LET*). This is just the average amount of energy lost per unit of par-

Figure 3 A. DIRECT ACTION;
Electrons, X or Gamma Rays; Low LET

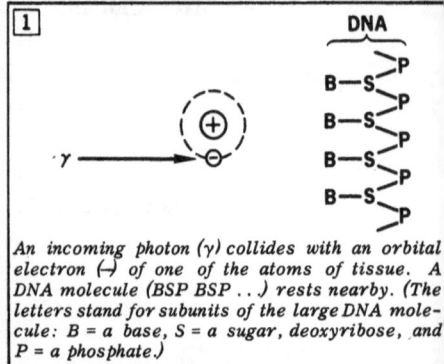

An incoming photon (γ) collides with an orbital electron (–) of one of the atoms of tissue. A DNA molecule (BSP BSP ..) rests nearby. (The letters stand for subunits of the large DNA molecule: B = a base, S = a sugar, deoxyribose, and P = a phosphate.)

Figure 3 B. INDIRECT ACTION: Electrons, X or Gamma Rays; Low LET

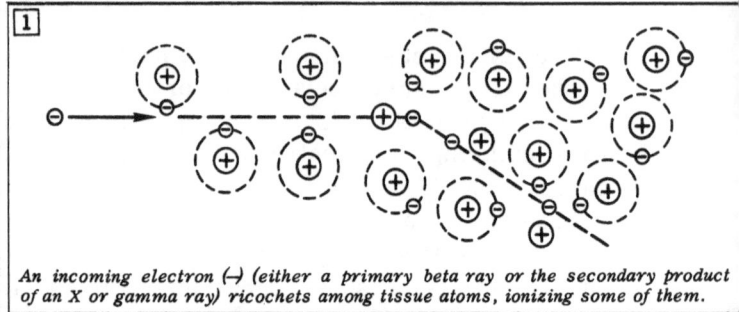

An incoming electron (–) (either a primary beta ray or the secondary product of an X or gamma ray) ricochets among tissue atoms, ionizing some of them.

Figure 3 C. INDIRECT ACTION: Protons or Fast Neutrons; High LET

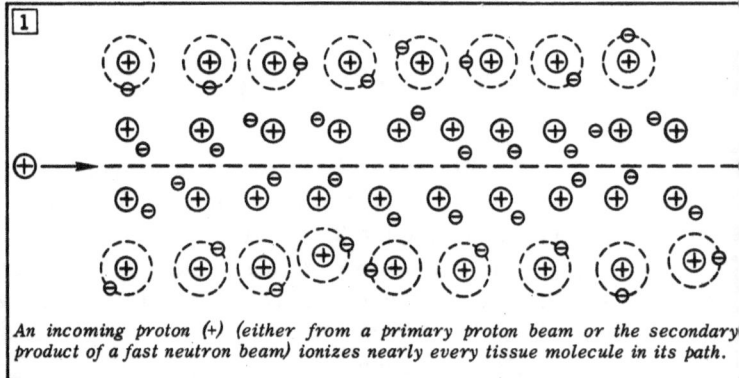

An incoming proton (+) (either from a primary proton beam or the secondary product of a fast neutron beam) ionizes nearly every tissue molecule in its path.

Figure 3 D. DIRECT ACTION:
Protons or Fast Neutrons; High LET

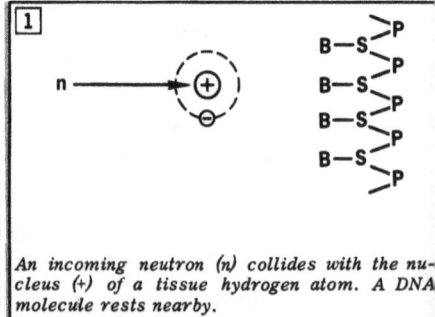

An incoming neutron (n) collides with the nucleus (+) of a tissue hydrogen atom. A DNA molecule rests nearby.

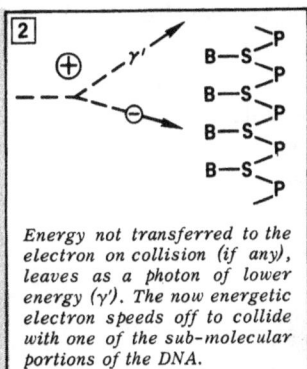

Energy not transferred to the electron on collision (if any), leaves as a photon of lower energy (γ'). The now energetic electron speeds off to collide with one of the sub-molecular portions of the DNA.

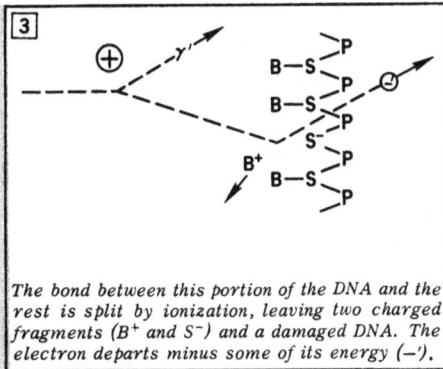

The bond between this portion of the DNA and the rest is split by ionization, leaving two charged fragments (B⁺ and S⁻) and a damaged DNA. The electron departs minus some of its energy (−').

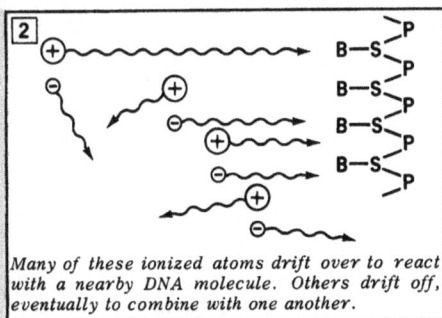

Many of these ionized atoms drift over to react with a nearby DNA molecule. Others drift off, eventually to combine with one another.

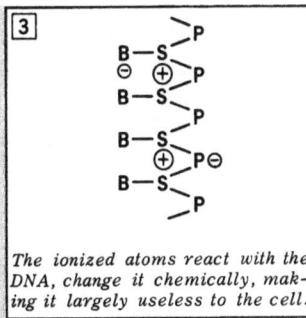

The ionized atoms react with the DNA, change it chemically, making it largely useless to the cell.

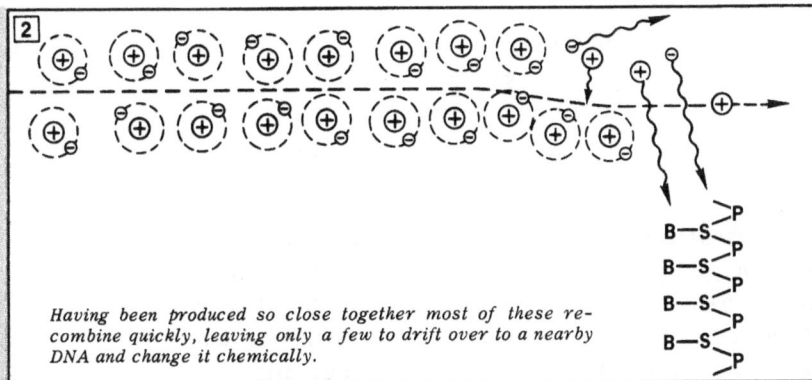

Having been produced so close together most of these recombine quickly, leaving only a few to drift over to a nearby DNA and change it chemically.

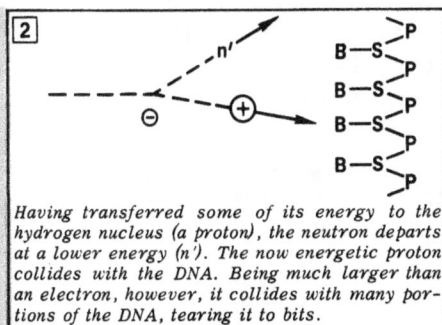

Having transferred some of its energy to the hydrogen nucleus (a proton), the neutron departs at a lower energy (n'). The now energetic proton collides with the DNA. Being much larger than an electron, however, it collides with many portions of the DNA, tearing it to bits.

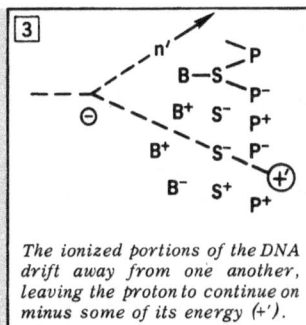

The ionized portions of the DNA drift away from one another, leaving the proton to continue on minus some of its energy (+').

ticle spur-track length. The average amount is specified, to even out the effect of a particle that is slowing down near the end of its path (2.42b) and to allow for the fact that secondary particles from photon or fast-neutron beams are not all of the same energy. Some average LET values are given in Figure 4, expressed in kev of energy lost per *micron** of spur-track length in tissue. As in the case of spur spacing, LET increases with increasing mass, increasing charge, and decreasing energy.

Figure 4   AVERAGE LET VALUES

Particle	Mass (amu*)	Charge	Energy (kev)	Average LET (kev/micron)	Tissue penetration (microns)
Electron	0.00055	−1	1	12.3	0.01
			10	2.3	1
			100	0.42	180
			1000	0.25	5000
Proton	1	+1	100	90	3
			2000	16	80
			5000	8	350
			10000	4	1400
			200000	0.7	300000
Deuteron	2	+1	10000	6	700
			200000	1.0	190000
Alpha	4	+2	100	260	1
			5000	95	35
			200000	5	20000

*Atomic mass units. On this scale one atom of normal hydrogen equals 1.008 amu.

For a first approximation, photons and neutrons may be thought of as producing electrons or protons, respectively, of about half the photon or neutron energy. The average LET of a 10-Mev fast-neutron beam would be about the same as that of a 5-Mev proton, or 8 kev per micron.

2.44   From our discussion of chemical, direct, and indirect effects (2.2, 2.3) we would expect the radicals and ions in the spurs of low-LET particles to diffuse away easily, and to react with any O_2 present to form $HO_2 \cdot$ radi-

*1 meter = 1000 mm = 1,000,000 microns (μ). Watt-fortnights per furlong might be substituted for those who prefer English units.

cals as in Figure 2E. From this we would predict that low-LET radiation would do most of its damage by indirect action, and that this damage would be greater in the presence of O_2 than in its absence. This, in fact, has been found experimentally to be the case for electron and photon radiation.

2.45   The interaction of a low-LET particle with a DNA molecule is shown in Figures 3A and 3B. The DNA is represented as a coiled chain of base, sugar, and phosphate groups. From Figure 3A we see that direct action damage is relatively small, since only one DNA bond is broken in the collision, and this can be "healed". The situation is analogous to one cigarette burn in a large blueprint: The missing lines easily can be drawn in again from those that remain.

From Figure 3B, on the other hand, we see that damage from indirect action can be much more serious. This is as if a large area of a blueprint had been charred by a blazing, oil-soaked rag. This would often, but not always, destroy an irreplaceable part of the print — one absolutely essential to future building with that part of the print (or to future cell growth and division in the case of DNA).

2.46   With high-LET radiation, spurs are so close together (Figure 3C) that recombination with neighboring spurs to reform water is very likely, and outward diffusion to react with tissue molecules or O_2 is unlikely. Indirect action is, accordingly, less important. It is as if a very hot, sharp welder's torch were held parallel to the blueprint. Some heat would reach and char the print but most would not.

On the other hand, if the torch were turned directly toward the print, the paper would be damaged immediately beyond all hope of repair. This is the direct-action situation (Figure 3D), and is the most important mode of action with high-LET particles. We would predict, from this, that damage to tissues from high-LET radiation would be almost impossible to repair and would be about the same in the absence of O_2 or its presence. This too has been proven experimentally to be the case.

2.47   In view of the fact that a high-LET particle that does not score a direct hit largely "wastes" its energy reshuffling water molecules, one might expect low-LET radiation to be the more damaging of the two, ev for ev. Experimentally, however, the reverse is true, and high-LET radiation is 2 to 20 times as damaging to tissues as low-LET radiation, ev for ev. Why?

The reason seems to lie in the amount of damage that a DNA molecule can sustain without losing its ability to repair itself. On the average, only 5 to 7 radicals from a low-LET spur can drift over to interact with one DNA, since the next spur will be so far away it will miss the DNA completely. Apparently DNA often can repair damage caused from this many "missiles".

With the blowtorch effect of the high-LET track, however, hundreds of sites on the DNA molecules are damaged and there is no hope of repair. A high-LET track may not strike through a DNA molecule as often as a low-LET spur, but when it does, the damage more than compensates for the rarity of the event.

2.5 Penetration

2.50 For charged particles the relationship between average LET and penetration is fairly obvious. The higher the LET (the more kev of energy lost per micron of travel) the sooner its energy is used up and the shorter its *range* (maximum penetration). Of course, LET itself drops with increasing energy, and a high-energy particle will have a greater range than would be predicted from simple proportion to a low-energy particle.

Of course, in reverse order, as a charged particle loses energy in tissue its LET increases, whereupon it loses energy even faster, whereupon its LET increases still more, and so on. This is the explanation for the rapid decrease in spur spacing toward the end of the tracks shown in 2.41. It also partly explains why we use average LET figures for comparing radiation biologically, but measured range for computing penetration.

The increase of ionization at the end of a charged-particle track* has been made use of in the treatment of

---

*Called the *Bragg effect*, after the modest British scientist, Sir William H. Bragg, who discovered it.

deep-seated tumors. Initially low-LET (by reason of high energy) charged particles will deposit a relatively small fraction of their energy near the skin and a larger fraction in the tumor. By the time they reach the tumor, they have lost enough energy so that their LET has increased.*

2.51 The uncharged particles (photons and neutrons) behave a little differently. Their progress through tissue resembles the path of a blindfolded man starting down a gravelly hill at a dead run. Most of the noise (ionization) he generates is made by the stones (charged particles) he kicks as he stumbles about. Bit by bit he tires and slows down. But he really hasn't a definite range — he just wanders on until he falls into a hole too deep to climb out of with the energy he has left (see Section 2.11). In the same way, by collision, photons generate electrons (pebbles) and neutrons generate protons (boulders), so that the average LET effects of the protons and neutrons are those of the particles generated. But the penetration of the uncharged particle itself remains much greater than that of charged particles of the same energy. Nevertheless, energy is still important, and a low-energy neutral particle (a weak man), while more penetrating than a charged particle of the same energy, is still less penetrating than if it had more energy (a strong man).

2.6 Amounts of Radiation

2.60 Like drugs, radiation can either heal or poison, depending on the amount given. Thus, amounts of radiation delivered are spoken of as *doses* and measurement of these amounts is called *dosimetry*.

The original international unit of radiation is called the *roentgen* (abbreviated r), after Wilhelm Roentgen, the discoverer of X rays. This unit depends on the ionization of air and, while useful for X and gamma rays, proved inadequate for the measurement of forms of radiation (such as neutrons) that ionize tissue better than they do air.

2.61 The most widely used unit today is the *rad* (an acronym for *r*adiation *a*bsorbed *d*ose). The rad is defined

---

*For more about radiation in diagnosis and treatment of disease, see *Radioisotopes in Medicine,* another booklet in this series.

as that quantity of radiation that delivers 100 ergs of energy to 1 gram of substance (tissue in our case). In practice the two units seldom differ by more than a few percent when we refer to tissue, so that we will consider them equivalent from here on.*

2.62 The *rem* (acronym for *r*oentgen *e*quivalent, *m*an) is a biological, rather than a physical, unit of radiation damage. It represents that quantity of radiation that is equivalent—in biological damage of a specified sort—to 1 rad of 250 *KVP†* X rays. If 1 rad of fast neutrons, for example, was found to lead to the death of as many rats as 3.2 rads of 250 KVP X rays, one rad of fast-neutron dose would be said to be equal to 3.2 rem for rat lethality.

The ratio of rem to rad is called the *relative biological effectiveness (RBE)* so that, in the example above, the RBE of fast neutrons would be said to be 3.2 for rat lethality. While the rem and RBE are related to the average LET for a given type of radiation, the numerical relation is not entirely clear. Hence, both rem and RBE comparisons are subject to greater errors than rad comparisons.

2.63 A teaspoonful of castor oil will have very different effects on a 25-gram mouse and a 70,000-gram man. What's important in dosimetry of any sort is not so much the total dose to the whole system as the dose-per-gram. That's why a physician prescribes different doses of medicine for different-sized people. The rad has already been defined as energy-per-gram to take this into account. Thus when we say that 1000 rads leads to the death of nearly any mammal we mean the delivery of $1000 \times 100 = 100,000$ ergs of radiation to each gram of the mammal's tissue.

2.64 Now 100,000 ergs of energy is not much by our usual standards. It's the energy delivered by a golf ball (44 grams) dropping $\frac{9}{10}$ of an inch (2.27 centimeters). Not even enough to raise a bruise! Dropping 70,000 of these, one for each gram of weight in the human body, would only

---

*Another unit, the *rep* may also be considered equivalent to the rad.

†Kilovolt Peak, a mixture of photons from 250 kev down to about 50 kev. When not otherwise noted, doses in this book will refer to this relatively standard radiation.

tickle. In terms of heat, 100,000 ergs = 0.0024 calories.* That much heat would only raise the temperature of a gram of tissue about 0.0025°C (0.0045°F). This is much less than the rise in body temperature brought about by a couple of deep breaths! Where, then, does radiation get its power to damage?

The answer, as we foresaw in 2.2, lies in its ability to change critical molecules, either by direct ionization or by indirect chemical reaction. Heat or mechanical energy is spread out more or less evenly among tissue molecules; no one molecule receives enough energy to injure it. But each particle of radiation packs enough "wallop" to smash to bits the chemical molecules it encounters. The situation is closer to that of small needles (0.01 grams) moving at the speed of pistol bullets (3000 centimeters per second). This energy is still only 100,000 ergs in each needle, but one of these penetrating each of the 70,000 cubic centimeters of material in a human body would be mighty damaging indeed!

2.7 Radiation Dosimetry

2.70 Since we cannot see, feel, or weigh radiation effectively (5.8), methods of radiation dosimetry depend on measuring its effects on matter. The common methods can be grouped into three basic sorts depending on whether the matter used is a gas, a liquid, or a solid.

2.71 The most common method simply measures the ionization produced in a gas. A typical device is shown diagrammatically in Figure 5. Radiation, striking the ionizable gas in the little chamber formed by the outer shield, causes ions to be formed—as we saw in Figure 2. If the voltage is high enough (a few hundred volts will do), the ions will be attracted to the shield or to the collecting electrode, depending on their charge, before they can recombine. As a result, a little "spike" (or *pulse*) of current will flow for each radiation particle that interacts in this way, and these pulses will be measured by the meter. When

---

*International calories, not nutritional Calories, which equal 1000 calories and should be spelled with a capital C.

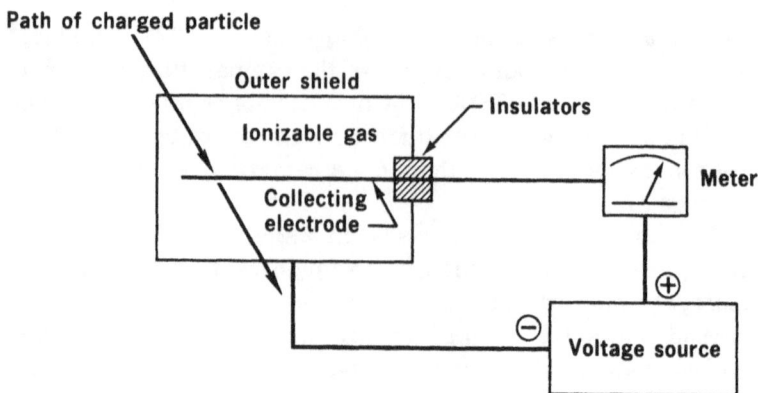

Path of charged particle

Outer shield

Ionizable gas

Insulators

Collecting electrode

Meter

Voltage source

Figure 5 *A typical gas ionization chamber.*

operated in this way, the device is called an *ion chamber*, and each pulse will be proportional to the ionization energy delivered by the particle.

If the voltage is increased somewhat, amplification occurs in the chamber and much larger pulses will pass through the meter. As long as these pulses are still proportional to the energy left by the particle, the device is called a *proportional counter*. Since the pulses are larger than before, small ionizations are more easily measured and a proportional counter is more sensitive than an ion chamber.

If the voltage is increased even further, the pulses become very large, but also become alike in size. This happens in the familiar *Geiger-Müller* counter. While very sensitive to the passage of radiation particles, it tells us little about their energy. As a result, it makes a rather crude dosimeter, but a very sensitive way of detecting small amounts of radiation.

Simple gases, like air or argon, are adequate for the dosimetry of charged particles (protons, electrons, etc.) or of photons (X or gamma rays). However, other types of radiation (such as neutrons) ionize tissue more readily than they do air or argon. In order to measure tissue doses for the many types of radiation encountered today, chambers have been developed with special gases, outer shields, and collecting electrodes that are equivalent to tissue in their composition. Such *tissue equivalent* ion chambers and pro-

18

portional counters are probably the most accurate and useful tissue dosimeters in use. Examples of different kinds of dosimeters are shown in Figure 6.

2.72   In liquid systems doses can be determined by measurement of the chemical changes brought about by radiation. Solutions of ferrous ion,* or of ceric ion,† in dilute sulfuric acid have proven quite useful in tissue dosimetry. Such a solution is mostly water, as is tissue. So conversion of measured dose to tissue dose is relatively simple. Since a liquid fits any container nicely, the use of shapes and sizes that correspond to men or laboratory animals is simplified. Measurement is made of the color changes that take place as a result of irradiation. Thus,

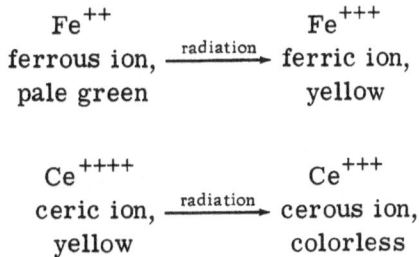

$$
\begin{array}{ccc}
Fe^{++} & & Fe^{+++} \\
\text{ferrous ion,} & \xrightarrow{\text{radiation}} & \text{ferric ion,} \\
\text{pale green} & & \text{yellow}
\end{array}
$$

$$
\begin{array}{ccc}
Ce^{++++} & & Ce^{+++} \\
\text{ceric ion,} & \xrightarrow{\text{radiation}} & \text{cerous ion,} \\
\text{yellow} & & \text{colorless}
\end{array}
$$

---

*Bivalent iron ions.   †Tetravalent cerium ions.

Figure 6 *Radiation dosimeters. 1. Air ionization chamber. 2. Tissue equivalent ionization chamber. 3. Meter for reading 1 and 2. 4. Small tissue equivalent chamber with tiny readable meter inside, to be carried on the person. 5. A portable Geiger-Müller counter including chamber (left) with meter and batteries. 6. A portable "Juno" air ionization chamber with meter. 7. A portable proportional counter. 8. A lithium fluoride luminescent dosimeter. 9. A chlorinated hydrocarbon liquid dosimeter. 10. A dosimeter containing packets of film sensitive to different types of radiation, to be carried on the person.*

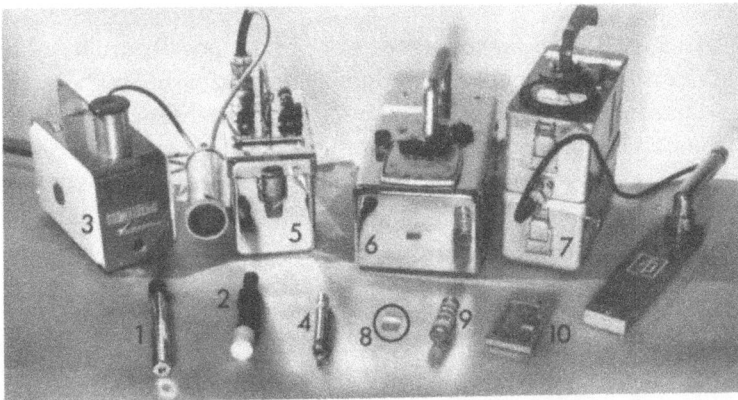

Although the smallest dose measurable this way is a few thousand rads, research promises to bring this down to 10 rads or so in the near future.

Another liquid system involves measurement of the hydrochloric acid freed from chlorinated hydrocarbons (for instance chloroform) by radiation. The color change in an indicator dye (such as litmus) can be related to the dose. Although maximum sensitivity is high (about 25 rads), chlorinated hydrocarbons are very poor equivalents for tissue.

2.73 The most ancient solid dosimeter of all, the gelatin emulsion photographic film, is still one of the most useful. It was originally used in the discovery of radioactivity by Henri Becquerel and for X-ray photography by Roentgen. Since their time special films have been developed for photons, neutrons, and charged particles. Although not as accurate a measure as many others, film is both sensitive and cheap, as well as roughly tissue equivalent.

In recent years some other solids (lithium fluoride, LiF, calcium fluoride, CaF_2, and some special glasses) have come into use. These solids can be made to give off light after irradiation and the amount of light measured is proportional to the radiation. Although sensitive, these solids are not, of course, tissue equivalent.

2.8 Radioactivity

2.80 Using a chunk of uranium ore and a few photographic plates Becquerel discovered a fact that was then very surprising: Not all the elements around us are the stable, workaday things people supposed. A few (chiefly thorium, Th, and uranium, U) are slowly but steadily breaking down, emitting alpha, beta, and gamma "rays", in their progress toward lead (Pb), the heaviest really stable element we know.

2.81 In time we came to know that not all atoms of an element are alike. The chemical properties of an atom are determined almost entirely by the numbers of protons in its nucleus. Atoms occur in nature, however, with the same number of protons (and, thus, the same chemical behavior) but different numbers of neutrons. These atoms, then, had different atomic weights* and could differ drastically in physical behavior.

Hydrogen atoms, for example, occur naturally in three sorts, all chemically the same, but differing in atomic weight and in physical properties. Such different physical forms of the same chemical element are called *isotopes* of one another. As a convenient shorthand, each isotope is designated by the symbol of the chemical element with its atomic mass (the total neutrons + protons) in the upper corner. Thus the three naturally occurring isotopes of hydrogen are ^{1}H, ^{2}H, and ^{3}H.†

2.82 The word isotope has often been used somewhat loosely to mean any single type of atom, rather than to refer to those atoms of differing atomic weights but the same number of protons. This is as if all atoms were to be called "hydrogens" rather than "atoms". The proper general term is *nuclide*. We'll use "nuclide" from here on, and save "isotope" to refer to the different nuclides of the same element.

The number of stable isotopes of a given element have been found to range from zero (technetium, Tc, promethium, Pm, and all the isotopes of elements with more protons than lead,‡ Pb) to nine (tin, Sn, and xenon, Xe). The number of unstable isotopes, on the other hand, is nearly unlimited. At least 900 unstable nuclides have been made or observed with the tools of physics, and many more remain to be discovered.

If, as is usually the case, an u n s t a b l e nuclide emits charged particles in its decay toward stability, the charge of the remaining nucleus changes, and therefore the chemical identity of the nucleus changes as well. One example of this *transmutation* of an element§ is given in 3.40. Another, example, of some importance biologically, is the radioac-

---

*Since protons and neutrons have nearly the same mass (or weight), but electrons only about $\frac{1}{1837}$ as much, the mass of an atom is governed almost completely by the total number of protons-plus-neutrons in its nucleus.

†Often named as protium, deuterium, and tritium from the Greek for 1, 2, and 3.

‡Bismuth, with 83 protons, actually is unstable, but its breakdown is so slow as to be nearly unobservable.

§ Dream of the ancient alchemists.

tive decay of one of the isotopes of natural potassium to give calcium:

$$^{40}_{19}\text{K} \rightarrow \,^{0}_{-1}\text{e} + \,^{40}_{20}\text{Ca}$$

2.83 Unstable nuclides differ widely in their stability, not only from nuclide to nuclide but also from individual atom to individual atom. However, for a given nuclide a very good sort of average life expectancy can be found, which is called the *half-life*. This is the time required, on the average, for half of the atoms of that nuclide to disintegrate. Thus if one starts at noon with 1 gram of ^{13}N (nitrogen-13), whose half-life is 10 minutes, one will have $\frac{1}{2}$ gram of radioactive material left at 12:10, $\frac{1}{4}$ at 12:20, $\frac{1}{8}$ at 12:30, $\frac{1}{16}$ at 12:40, etc. Eventually (about 2:30 AM) there will be only a single radioactive atom. Just when this one will disintegrate is hard to tell, but the odds are 50 : 50 that it will do so within the next 10 minutes. The situation is similar to that of a leaky tire. At first, because the tire pressure is high, many air molecules are compressed together near the hole and they escape busily with a happy hiss. As the pressure drops from leakage, fewer molecules are near the hole and they need to travel farther to reach it. Eventually only a few are left, and they drift "lazily" out of the now flat tire.

2.84 Radioactivity is measured in *curies;* one curie is the amount of any radioactive nuclide that decays at the rate of 3.7×10^{10} (37 billion) disintegrations per second. This *decay rate* is always inversely proportional to the half-life and directly proportional to the number of unstable atoms* present, so that one can always compute one if

---

*We say atoms, rather than nuclei, to include nuclides that decay by obligatory electron capture (such as ^{7}Be) and could not decay at all in the absence of orbital electrons. Applicable formulas are:

$$C = \frac{(0.693)(N_1)}{(T_{\frac{1}{2}})(3.7 \times 10^{10})} \quad \text{and} \quad N_2 = \frac{(N_1)}{2^{(T_{\frac{1}{2}})/(t_1-t_2)}}$$

where C is the decay rate in curies, $T_{\frac{1}{2}}$ the half-life in seconds and N_1 and N_2 the numbers of atoms at times t_1 and t_2. All times are expressed in seconds.

given the others. However, each radioactive nuclide decays by emitting its own characteristic pattern of alpha, beta, and gamma particles so that computations of the radiation dose delivered by a radioactive source are necessarily complex. We will in our discussion give the rad dose itself where it is important.

2.9 Radiation Protection

2.90   The best way to avoid becoming a weekend traffic statistic is to stay off the highways. With radiation, too, the best policy is avoidance. But avoidance involves time, space, distance, and the possibility of shielding. A little discussion of each is in order.

2.91   Damage from radiation tends to be largely cumulative* so that, in general, a few thousand rads will be lethal whether given over seconds or years. The maximum lifetime dose that can be considered "safe" has been set at about 250 rads. One could consider staying in a 20-rad-per hour field for a few hours if need arose, provided one had some certainty that the experience would not be repeated. On the other hand, if one's daily work involved exposure to radiation (as is the case for X-ray therapists, atomic energy workers, etc.) one would not care to tolerate steady fields of much more than a few *milli*rads per hour. Thus, time-avoidance is governed largely by consideration of the total dose that one is likely to accumulate over a lifetime.

2.92   Spatially one might tolerate a dose of several thousand rads to a limited part of the body in an emergency or for the cure of a serious disorder, where a whole body dose of this size would certainly be lethal. This amounts to tolerating a badly burnt hand to save the body, where the same burn over the whole body would be fatal.

2.93   In distance the most useful principle is the so-called *inverse-square law*. It states that the radiation dose rate from a small† source drops off as the square of the distance from the source. That is the dose at 2, 3, 4, 5, etc., meters from a small source will be $\frac{1}{4}$, $\frac{1}{9}$, $\frac{1}{16}$, $\frac{1}{25}$

---

*This is not always true, but we'll cover the exceptions in Section 4.4.

†The law applies well at distances 5 or more times the diameter of the source.

etc., of the dose at 1 meter. This behavior, of course, is that seen with most radiating bodies — the light intensity from a light bulb, the heat from a glowing coal, and so on.

The reasoning is simple enough. Since the radiation is emitted spherically (equally in all directions*), its intensity decreases with increasing distance by the necessity of "smearing" itself equally over the whole area available. Since this area increases as the square of the distance,† the intensity must decrease in proportion. Polka dots on a balloon get four times their original size when the balloon is blown up to twice its previous diameter.

2.94 *Shielding* against charged particles is relatively easy, since they interact so well with matter. Thin sheets of metal, paper, or even the air itself often provide nearly complete shielding.

The uncharged X-ray and gamma photons, and neutrons, tell us a different story. Lead is about as opaque to gamma rays as Kleenex is to light. Because complete absorption is unlikely, shielding materials are compared in terms of their *half-value thickness* for a given radiation. This is the thickness that will reduce the radiation intensity to one-half its unshielded value. The same sort of law applies as for half-lives; two thicknesses yield $\frac{1}{4}$, three $\frac{1}{8}$, four $\frac{1}{16}$, five $\frac{1}{32}$ the dose, etc. The half-value thickness of lead for cobalt-60, ^{60}Co, gamma rays is about 1 centimeter, for example, while the half-value thicknesses of earth and water are about 5 centimeters. Earth and water often are used if space permits because they are inexpensive.

After about 10 half-value lengths, there are buildup and backscatter effects that occur and alter the computations some, but the method gives a good idea of the amount of shielding needed even so.

---

*A beam of radiation is an obvious exception. But, then, one just steps aside to avoid a beam.

†The area of the surface of a sphere being $4\pi r^2$.

3 BACKGROUND AND FOREGROUND—SOURCES OF RADIATION

3.1 Natural Backgrounds

3.10   The old saw tells us that "into each life some rain must fall", but it omits mention of radiation, which is even more widespread than rainfall. The *external radiation background* consists almost entirely of *cosmic rays** and of gamma rays from the natural potassium-40, ^{40}K, thorium-232, ^{232}Th, and uranium† of the soil. The charged particles from the latter, and from the small amounts of ^{14}C and ^{3}H constantly being produced on earth by cosmic-ray action, penetrate air and skin so poorly that they add little to the natural external background.

3.11   An *internal background* also exists, since each of the naturally radioactive nuclides finds its way into the body. The most important, of course, are ^{40}K, ^{14}C, and ^{3}H since our bodies contain relatively high percentages of potassium, carbon, and hydrogen. Thorium and uranium also enter the body and, as they decay, pass through a number of radioactive daughter elements before eventually becoming lead. Like Th and U themselves most of these daughters pass through the body rapidly. But at least one, *radium*, Ra, chemically resembles natural calcium so much that it tends to pile up in our bones (see Figure 7). In this way it, too, adds noticeably to our internal background.

All in all, these sources bring our natural radiation dose to about 0.15 rad per year or less, for a lifetime dose of about 10 rads. A few places on earth, notably in India and Brazil, have soils so rich in Th and U that the natural background may be as high as 2 rads per year. Also, air tends to act as a shield against cosmic-ray dose so that "dwellers in high places" (such as mountainous regions) may receive a little more radiation than those at sea level.

---

*These "rays" consist of particles and photons of extremely high energies and of every conceivable sort, whose source is still not clear. Included here are high-energy radiations from our sun, since it is very difficult to separate the two. See Section 3.6.

†A number of other natural radioactive sources are not mentioned here, but their contribution is negligible.

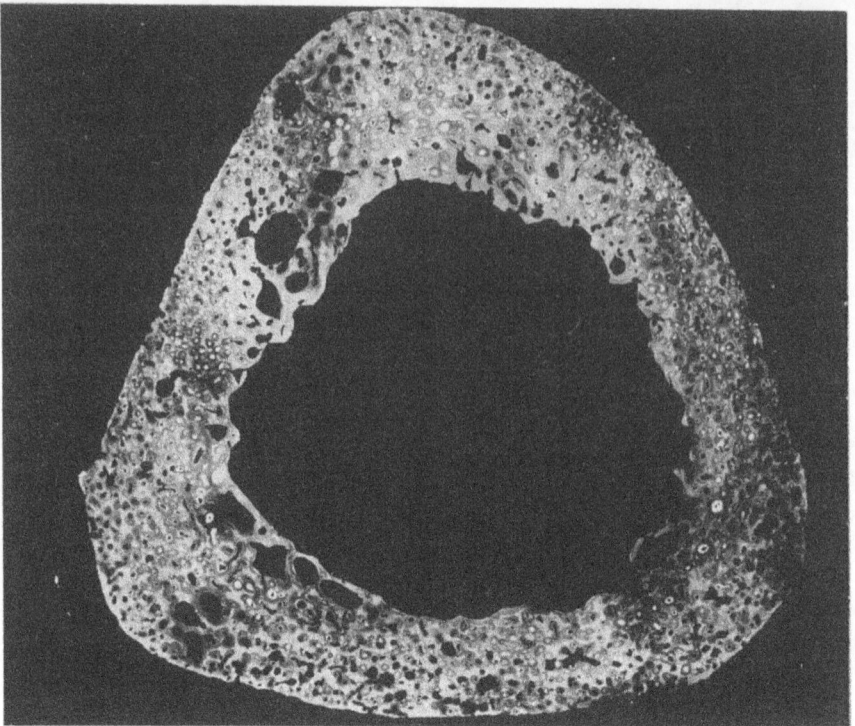

Figure 7 *A section of bone from the body of former radium watch-dial painter, who, in order to maintain a fine brush tip, was in the habit of touching the tip with his tongue. The photograph on the left shows darkened areas of damaged bone. On the right is an autoradiograph, in which*

But 0.15 rad per year is a safe maximum figure for most of us. A breakdown for residents of the United States is shown in Figure 8.

Figure 8   THE NATURAL BACKGROUND

Type	Source	Dose rate in millirads/yr	Varies with
External	Cosmic rays	30-60	Latitude and altitude
	Soil potassium-40, thorium, uranium	30-100	Location (mineral deposits) and dwelling (least in tents, greatest in stone buildings)
Internal	Thorium, uranium and daughters	40-400	Location and water supply
	Potassium-40	20	Not very variable
	Carbon-14	2	Not very variable
	Tritium	2	Not very variable
Total		100-600	

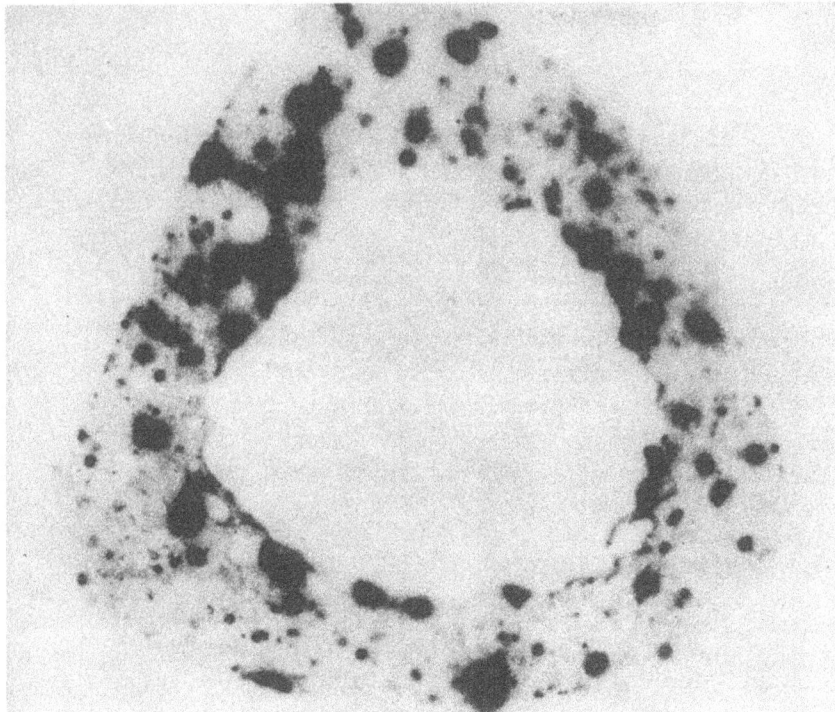

the bone "took its own picture" by being held against film, showing areas exposed by the radium alpha particles. Note the areas of high alpha activity correspond to the areas of maximum damage in the photograph (left).

3.2 Medical Exposures

3.20 Radiation and radioactivity, in the form of radioactive nuclides, X and gamma rays, are unquestionably among medicine's most useful tools today. Like all powerful tools, however, they must be used carefully. Dosage to the patient varies widely with the type of radiation and the application. But nowadays* one can be certain that the unavoidable medical "background" will not exceed the minimum needed for effective medical use.

At one extreme lies the patient who may need 6000 rads for the treatment of a tumor that would otherwise be fatal. Of course such a dose is concentrated at or near the tumor so that the average whole body dose is much less. Still,

---

*This was not always true, since radiation hazards were not as well understood in the past as they are today. As little as a decade ago a few routine medical and dental X-ray examinations might deliver perhaps 300 rads locally. Newer methods have largely eliminated this.

even an average body dose of 200 rads would be too high to contemplate in anything but a life-or-death situation. At the other extreme some people go through life never having even a dental or a chest X ray.

3.21   The average American has about four dental X rays and one medical X ray every 10 years. With modern equipment and practice this will result in perhaps 1 rad per year of local dose, but only 50 millirads or so of whole body dose. All in all, radiation from medical sources exceeds the natural background for only a few of us. Nevertheless, research is ever in progress to find ways of lessening the dose needed.

3.3 Artificial Backgrounds

3.30   A number of objects of our daily experience constitute sources of noticeable radiation. A television set, for example, is simply a low voltage X-ray machine. Electrons from the picture tube filament are speeded up by electrical voltage until they collide with the phosphor coating of the tube. This is precisely the mode of operation of an X-ray tube. However, the low voltages employed, the poor efficiency of the phosphor for X-ray production, the glass and plastic shielding through which the X rays must pass and the distance of the average viewer from the screen all combine to lower the received dose. On the average this source contributes less than one millirad per year to one's total dose.

3.31   For many years various radium-containing paints have been used to render watch and clock dials luminous. Since radium and its daughters emit both gamma and beta rays energetic enough to penetrate a watch glass, the dose rates from watch faces may exceed 2 millirads per hour and a clock face 200. Depending on one's habits these sources may contribute 10 to 1000 millirads per year to one's total dose; 25 is probably a good average figure for a person who uses both one watch and one clock of this sort. In recent years ^{3}H has come into vogue for these self-luminescent paints.* It emits only beta rays of energies low enough

---

*^{90}Sr was used for a time but has been replaced by ^{3}H whose betas are even less penetrating.

to be completely stopped by glass or plastic, so the dose from a watch with this material on the face is about nil.

3.32 A few other scattered sources of radiation turn up at times. The bathtub sets glazed with a uranium pigment (uranium compounds have a lovely yellow color) that afforded the users a dose rate of over 100 millirads per hour might be mentioned. Or the houses built of radioactive stone that bombarded occupants with 10 millirads per hour. But such sources really constitute accidents rather than environments and contribute little to the average dose.

3.4 Fallout*

3.40 When a nuclear weapon is detonated a number of radioactive nuclides are produced that may slowly come to earth far from the site of the detonation. The nuclides of chief interest biologically arise in three ways. Atoms of

---

*For a more extensive treatment see the booklet *Fallout from Nuclear Tests* of this series.

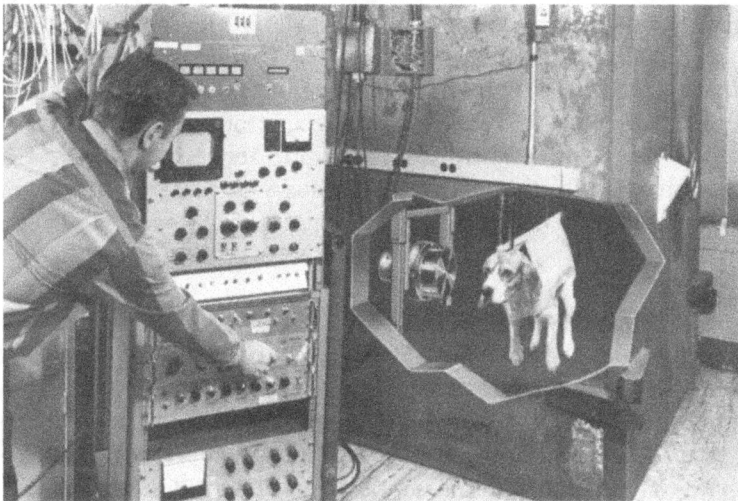

Figure 9 *Measurement of fallout nuclides in the body of a dog. Gamma rays are detected by a sodium iodide crystal, which records them as electrical pulses. An electronic analyzer gives a spectrum of the number and kinds of gammas present, and from this the natural and fallout nuclides in the body can be determined. External background is diminished by a shield (cut away to show the dog inside).*

plutonium-239 (^{239}Pu) from the w e a p o n's core may be scattered by the blast, as may cesium-137 (^{137}Cs), strontium-90 (^{90}Sr), strontium-89 (^{89}Sr), and iodine-131 (^{131}I) formed by the fission of ^{239}Pu itself. Also, the neutrons produced by the weapon may be absorbed by the nitrogen-14 (^{14}N) of the air to give carbon-14 (^{14}C), according to the equation

$$^{14}N + n^1 \rightarrow {}^{14}C + {}^1H$$

One atom of nitrogen-14, in other words, captures a neutron, then liberates a proton to become carbon-14.

3.41 The average dose from fallout in the United States is less than 15 millirads per year—less than 10% of the natural background. The chief danger from fallout lies not so much in its total dose as in the fact that many of the nuclides resemble normal body elements in their chemical behavior. Thus, as with the radium discussed in section 3.1, each may find its way to a *critical organ* or *tissue* (5.50) and exert an effect out of proportion to its average dose. ^{89}Sr and ^{90}Sr resemble calcium and, so, will find their way to the bones. ^{131}I will behave like stable iodine* and concentrate in the thyroid and salivary glands. ^{239}Pu finds its way to certain sensitive sites in bone. ^{14}C, following normal carbon,† may lodge within the critical DNA molecules of our cell nuclei. Despite these localizations it is improbable that the total damage to be expected from fallout approaches the level of that from the natural background.

3.5 Industrial and Scientific Sources

3.50 As the industrial and scientific use of radiation increases, the number and type of sources used increases also. At present a list of these would at least include nuclear reactors, high-energy particle accelerators, industrial X-ray and gamma-ray sources for materials inspection, irradiators for the sterilization and preservation of food, radioactive nuclides used for tracer studies, alpha-ray, beta-ray, gamma-ray, X-ray and even neutron sources

---

*Which happens to be all ^{127}I.
†Which is 98.9% ^{12}C, and 1.1% ^{13}C, and a smidgen of ^{14}C.

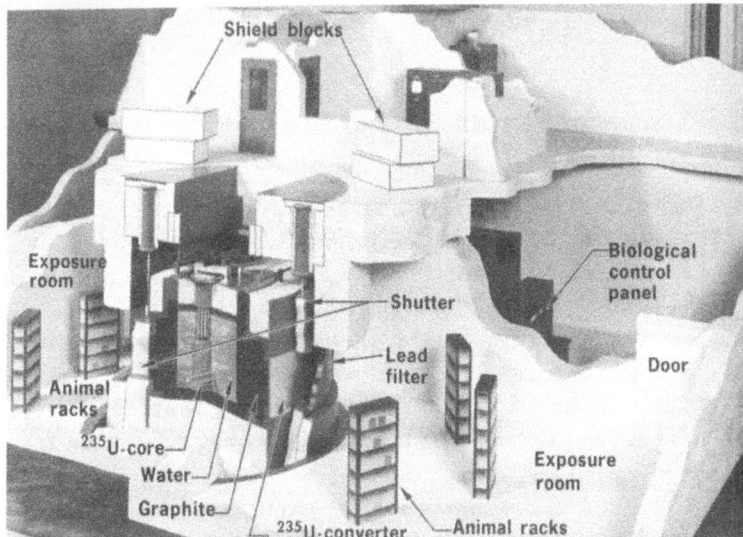

Figure 10 *A model of the JANUS Biological Reactor at the Argonne National Laboratory. Neutrons produced by the fission of ^{235}U in the core are slowed and reflected back to the core, by the water and graphite surrounding it, to sustain the chain reaction. Excess slow neutrons may be used immediately or changed to fast fission neutrons by the ^{235}U converter. They then pass through the lead filter, which removes gamma rays, into the exposure rooms—one on each side of the reactor (hence the name JANUS, after the two-faced Roman god). Doses and dose rates are controlled from the biological control panels, by raising and lowering the shutters, and by changing reactor power.*

for industrial product analysis, measurement, and control, and irradiation cells used for specialized chemical reactions such as the production of toughened plastics.* Despite widespread use of these sources, safety controls are so rigorously maintained† that they contribute nothing to the general radiation level and are not likely to do so in the future. Workers at these facilities, of course, must be prepared to handle the possibility of undesirable exposure as in any other occupation involving potential hazard. As a guide line the International Commission on Radiation Protection (ICRP) has advised that total exposures averaging over 100 millirads per week (5 rads per year) be avoided. In practice safety measures have been so effective that few workers even approach this level.

---

*See *Radioisotopes in Industry*, another booklet in this series, for a discussion of these uses.

†Partly by the USAEC and other federal and state agencies, and also by the users themselves.

3.6 Sputniks, Spacemen, and Speculation

3.60   As man prepares to wander off to the stars, new (and presumably exciting) sources of radiation await discovery. Although present knowledge is far from complete, we are aware of at least three sources of radiation in space.

3.61   The most obvious, of course, are the highly penetrating cosmic rays whose energies run up to 10^{18} ev or more. Our air blanket provides a partial shield here on earth so that instead of the 50 millirads per year we receive at sea level we will find about 10 rads per year in free space.

3.62   Next come the various particles emitted by the sun, or any other star, for that matter. These consist chiefly of high-energy protons (up to 1000 Mev) and lower energy electrons (up to 2 Mev). Their intensity increases considerably during solar flares and other "storms" on the sun. But the overall dose to man in a space capsule from this source would probably not exceed 100 rads during a 1 year mission. Of course the inverse-square law still applies so that this dose drops off as one leaves the earth headed away from the sun, for example, toward Mars.

3.63   Right around earth, the famous *Van Allen Belts* make things a little hotter. They consist chiefly of protons and electrons trapped by the earth's magnetic field. The dose in these belts can rise to perhaps 1000 rads per day but they are only a few hundred miles deep so that space travelers will simply avoid spending much time in them. Their proton composition is largely that of the solar radiation, since that was their source. The electrons, however, have energies up to 8 Mev or so, the ones with energies over 2 Mev having been contributed by a nuclear test explosion in space on July 9, 1962.

Figure 11   *Normal cell division. 1. A metabolic cell. 2. The chromosomes duplicate and the nuclear membrane disappears. 3. The sets of chromo-*

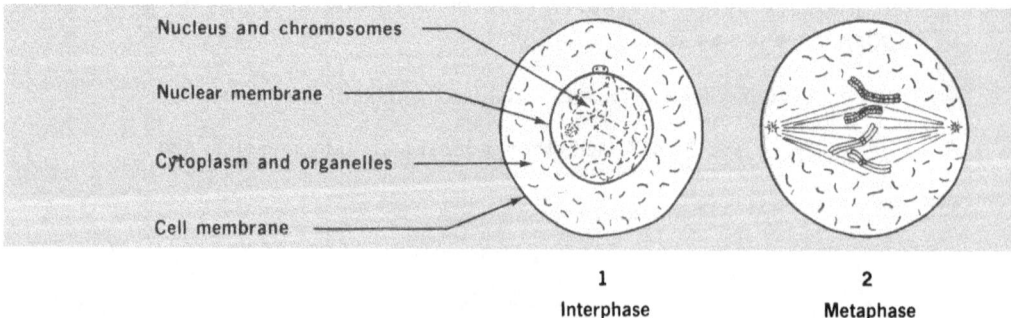

Nucleus and chromosomes

Nuclear membrane

Cytoplasm and organelles

Cell membrane

1
Interphase

2
Metaphase

All in all, with adequate shielding and a quick scooting through the Van Allen Belts the total dose to astronauts can probably be kept well below 150 rads per year. What levels of radiation voyagers to the far stars will find lies still hidden "in the womb of time".

4 OF MICE AND MEN, OR SOME BIOLOGY AND MEDICINE

4.1 The Cell

4.10   The cell is, of course, the fundamental unit of life as we know it. It also appears to be the most important single entity in radiobiology as well. Now this might seem too obvious to mention, but when these matters were first studied, the possibility existed that the chief effect of radiation would be the disruption of the organization or communication system of the body, which would leave the cells relatively uninjured. Considerable evidence has accumulated, however, to indicate that the cell, and particularly its nucleus, is the primary site of radiation damage. Disorganization of the body follows only after a sufficient number of cells have been so injured that they can no longer carry out their normal functions.

4.11   The average cell lifetime in a body is much less than that of the body itself, just as individual lifetimes in a human society are much less than that of the society itself. Accordingly one of the most important functions of a cell is its own reproduction. Cells reproduce themselves normally by a process of cell division called *mitosis*. As a cell dies it is replaced by the daughters of its sister cells. A simplified diagram of a mammalian cell undergoing mitosis is given in Figure 11.

somes, one destined for each new cell, move apart. 4. The cytoplasm begins to divide and the nuclear membrane reforms. 5. Two new metabolic cells.

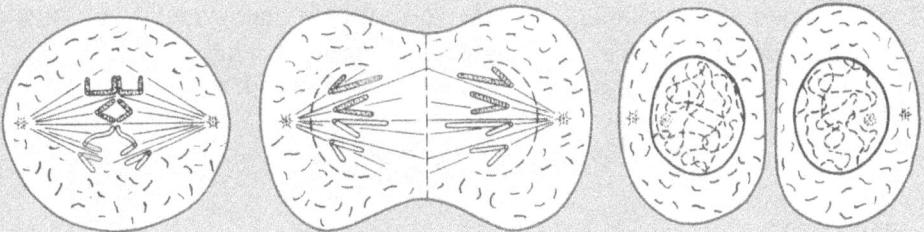

3	4	5
Anaphase	Telophase	Interphase, daughter cells

4.12   The *cell membrane* encloses the *cytoplasm*. Each of these is in a state of constant physicochemical activity called *metabolism*. The cell membrane controls the entrance and exit of water, ions, nutrients, and waste products by a complex chain of biochemical and biophysical reactions. The cytoplasm, aided by the various *organelles** (little organs) within it, carries on the main metabolic activity of the cell. Within the cytoplasm lies the *cell nucleus*, a volume of chemically distinct *nucleoplasm* encompassed by its own *nuclear membrane*. It shares the cytoplasm's metabolic activity, but in a different way. The chief function of the nucleus appears to be that of overall control of the cell. It acts as an executive whose records and blueprints are contained in the biochemical coding of its DNA molecules (2.3), which are organized into nuclear organelles called *chromosomes*. The very ability of the cell to repair itself depends on its ability to read these blueprints. Their destruction or damage leaves the cell without proper reference material on which to base needed biochemical "decisions".

4.13   If damage to the coding on a DNA molecule is slight and occurs during that part of a cell's lifetime when it contains two sets of DNA, the damage can apparently be repaired properly and the cell returns to normal (see Figure 12). If the damage is more severe and leaves the cell with one or more destroyed blueprints, it may be able to maintain metabolic activity but loses its previous identity. This cellular "amnesia" may render the cell unfit to continue in its environment, and it or its daughters may eventually die away.

However, the "amnesic" cell may be quite vigorous in its environment and yet have forgotten its previous identity so completely as to have passed out of the growth control of the body. All normal cells are controlled in their growth by the needs and demands of the body as a whole. But this susceptibility to control apparently requires a specific normal complement of DNA molecules in the nucleus.

For an analogy we might consider that control of army groups by their general requires working radio, telephones,

---

*Mitochondria, microsomes, Golgi apparatus, vacuoles, etc.

34

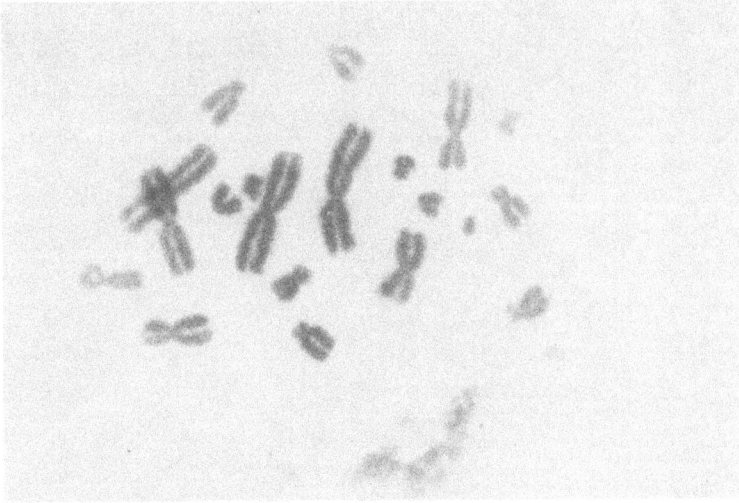

Figure 12 *This normal cell nucleus, enlarged 550 times, has 23 pairs of chromosomes.*

mail, messenger service, etc., as well as loyalty. If these break down within a given group that group passes out of control. Such a unit might be lost for effective use to the general, or it might attempt to gain control of the entire army. In the case of the body such an uncontrolled cell becomes a *cancer cell**(see Figure 13). As with the army group, it may die away, it may simply remain —growing slowly and doing little damage (a benign tumor), or it may invade and destroy its host (a malignant cancer).

4.14   If damage is still more severe the cell will lose its ability to divide properly and will die, usually after a period of confused growth of *giant cells* (see Figure 14). If damage is more severe than this, not only DNA but other cell components are damaged beyond repair. Cellular activity slows markedly, the cell becomes visibly abnormal, and it dies quickly.

4.15   Radiobiological damage is considered to occur, then, in this light. If a critical DNA molecule in a cell nucleus is damaged by radiation, the cell becomes deranged. If enough cells become deranged, the tissue, then

---

*This picture, albeit oversimplified, is remarkably useful and nearly "true".

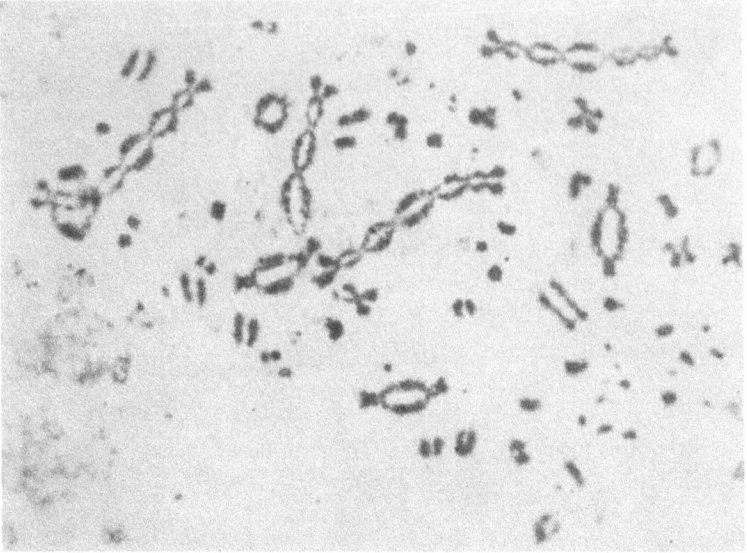

Figure 13 *This cell nucleus, enlarged 275 times, has chromosomes that have doubled, tripled, etc., abnormally because of irradiation. Such a cell could easily become cancerous.*

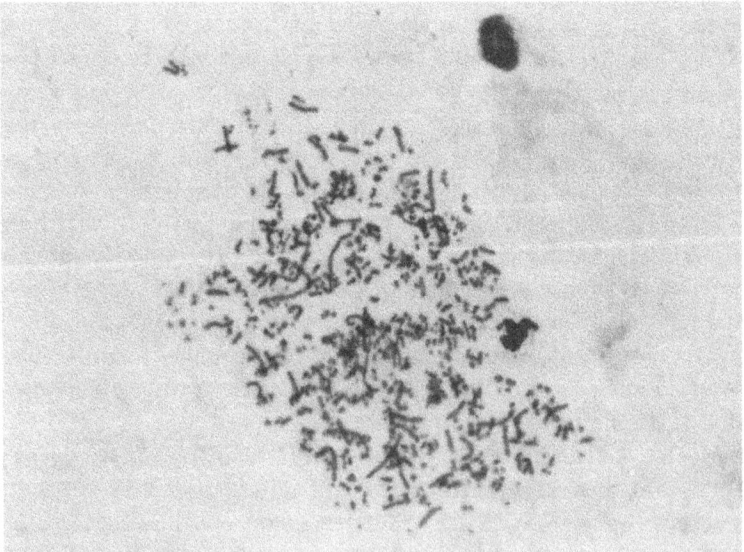

Figure 14 *This giant cell nucleus, enlarged 50 times, has grown abnormally due to irradiation. The cell now has more than 700 chromosomes and is about 10 times the size of a normal cell.*

the organ, and finally the body* become disordered to a degree that depends on the severity of the damage to the DNA molecules and the number and relative importance of the cells deranged. Thus radiation damage ranges from the unimportant death of a few replaceable cells (much like a mild sunburn) through the induction of tumors and cancers, to premature aging, acute illness (radiation sickness), and to nearly immediate death.

4.16   A few human cells, notably those of the nerves (*neurones*) and the red cells of the blood (*erythrocytes*), are incapable of division. The latter cannot divide because they have no nucleus. But, as they die in the normal course of events,† new ones are quickly supplied by the *hematopoietic* (blood-forming) *systems* of the spleen and bone marrow. *Nerve cells*, on the other hand, have nuclei and are able to regenerate a lost fibre (*axone*) or other cell part. But the nuclei themselves simply seem to be incapable of division; thus when once destroyed, they can never be supplied.

4.17   Two aspects of cell death from such nucleoplasmic poisons as radiation should be stressed. The first is that the low *redundancy* (the low degree of repetition of identical molecules) in the nucleus makes a little damage there much more serious than the same amount of damage in the much more redundant cytoplasm. One can destroy a great many tools on a job without slowing it down much. But one important blueprint destroyed will, eventually, stop the whole project.‡

4.18   The other aspect is the delay between nucleoplasmic damage and obvious derangement of the rest of the cell. The destruction of an important blueprint is not noticed until it is needed. Similarly damage to the nucleus isn't obvious until the cell tries to divide!

4.19   But, by the same token, much more radiation§ is needed to kill the cell outright than to damage the nucleus

---

*"For want of a nail, ..."

†Their average life span is about 120 days.

‡Unlike the case on a construction project, the chief architect of life is not available for reconsultation.

§About 10,000 times as much. That's why radiation "death rays" have never been very practical. A hand weapon is not much use if your adversary won't feel its effects until next Tuesday.

badly enough to ensure the cell's death at its next division. This is simply because the most important and easily damaged molecules of the cytoplasm, probably its *enzymes* (the protein catalysts responsible for the cell metabolism) occur in relative profusion. Nearly all the many enzyme molecules of a particular kind must be destroyed before the cell dies for lack of them.

4.2 Poisons and People

4.20 If radiation can be understood in terms of its ability to poison cell nuclei, a little discussion of *toxicology* (the study of poisons) and of human medicine is in order. Radiation, as a poison, has some effects that are similar to those of the lead salts. It can be delivered all over the body at once; it can be delivered to certain parts of the body; its effects are cumulative; and some body tissues are more affected than others.

4.21 Also, like insoluble lead salts, radiation effects can be *abscopal.** This means that the results of local damage can be felt elsewhere, usually in a different form. A bad burn on the leg, for another example, can give one a bad headache as well as a feeling of being "miserable all over".

4.22 In three respects radiation, as a poison, is nearly† unique. The first lies in its nearly specific effect on the critical molecules of the nucleoplasm (see 4.1). Few chemicals can behave in this way, because they must diffuse slowly through the cytoplasm before they can reach the nucleus and are usually "captured" on their way through.

Second, the damage is done at high speeds. Unlike chemical poisons radiation cannot be removed before its full damage is complete. It's literally "gone before you know it". Thus the only feasible answers are: Avoid it, put molecules in its path that will be sacrificed to save those of the nucleoplasm, or attempt to repair the damage itself once it has occurred.

Finally radiation can destroy the body's normal *immune response*. A body's best defense against outside invaders

---

*Literally "away from the place you're watching".

†Certain chemical poisons are said to be *radiomimetic*, because they mimic the actions of radiation.

is its ability to rapidly develop specific biochemical defenses, called *antibodies*, that act against the invaders. Many antibodies continue in production for life once they begin, and make one immune to further attack by that particular invader.* Others last only as long as the invasion is active. Radiation can destroy the body's ability to produce or maintain these antibodies and, in this way, leaves one prey to serious infection by any invader that comes along.

4.23 Now toxicology, especially that of the nucleoplasm, would be greatly simplified if one could see the chemical reactions of the cell. But this is impossible in practice so that a physician must rely on the *symptoms* (observable special effects) of poisoning and make his diagnosis on the basis of how well these symptoms fit a given *syndrome* (a symptom group characteristic of a certain disorder). Nausea, vomiting, hemorrhage, diarrhea, loss of weight, and severe anemia are symptoms that, taken together with the possibility of exposure to radiation, constitute the syndrome of acute radiation illness. In the same way cough, headache, fever, and a runny nose are the symptoms that, taken together with "winter weather and runny-nosed friends", add up to the syndrome of a bad cold.†

4.3 The Laws of Averages

4.30 The severity of the symptoms of poisoning will depend on the dose of course. But it will also depend very much on the individual as well. Even identical twins will differ in their responses to a given situation. Accordingly, one is reduced to studying the reactions of groups large enough for the laws of statistics to apply. In this way one hopes to derive useful averages, which are applicable with fair certainty to human populations.

4.31 Except for Hiroshima, Nagasaki, a few accidents, and patients undergoing radiation treatments we have had (fortunately!) little opportunity to study the effects of radiation on people. Most of our information has been obtained from mice and other happily cooperative laboratory

---

*Mumps, for example.
†Coryza, for collectors of jargon.

animals. This information, extended to humans by the laws of scaling,* provides us with the best averages we have at the moment. The success that this approach has had in predicting the effects of foods, drugs, anesthetics, and poisons in humans gives us a certain confidence in our results.

4.32 The sort of individual response one finds in a population of animals to different doses of radiation† is shown in Figure 15. As the dose is increased from zero, very few animals show any effect at first, then more and more do, until 50% of the animals are affected. At 50% the percentage being affected begins to drop off‡ until greater and greater doses are needed to affect the remaining few. At last even the "super mice" succumb and 100% are affected.

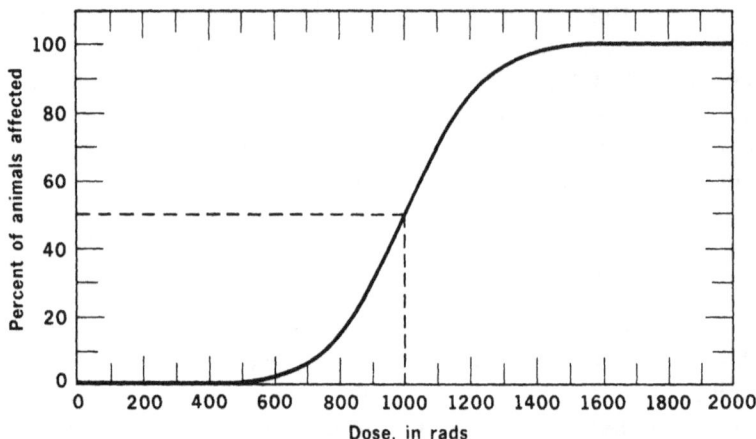

Figure 15 *A dose-effect distribution.*

Although the curve§ given is that for the death of mice exposed to ^{60}Co gamma rays, similar curves of the same sort apply equally well to other *dose-effect* studies. Thus,

---

*That is, according to the increase in scale, or relative size, of human beings compared to the experimental organism.

†Or of nearly any other poison for that matter.

‡We say that there is an inflection (turnaround) at 50%.

§This type of curve is called "sigmoid" from the Greek for the letter S, because its shape is that of a rather lazy S.

with a change of dose scale, it would fit for the percent of humans cured of infection by increasing doses of penicillin or the number of little boys developing tummy aches on eating increasing numbers of green apples.

4.33 Since the largest percentage of individuals will be affected by the smallest change in dose* at the 50% level we ordinarily use this point for reference. Thus a *D-50 value* is that dose at which 50% of the population studied showed the effect described. If the effect is death the symbol *LD-50* (lethal dose-50%) is used, if some other effect *ED-50* (effective dose-50%).

4.34 Now not all effects will be noticeable in the same time periods. It may take a larger dose of green apples to give 50% of the boys tummy aches during an afternoon than it will by school time the next morning. Therefore symbols like *LD-50/30* (the dose that kills 50% within 30 days) are used.

4.4 Doses, Dose Rates, and Recovery

4.40 Just how nasty a poison can be depends, of course, on its natural nastiness (LET) and on the total dose (rads). If living creatures weren't (living, that is), that would be all there was to it. But cells and people alike are able to recover from unpleasantness, the degree of *recovery* depending not only on the unpleasantness itself but also on the time between successive doses.†

4.41 Radiation can occur steadily or in intermittent chunks. Furthermore, it can be delivered to the whole body or to parts of it. Each of these factors will have its own effect.

If the radiation occurs steadily, the situation is similar to that of Napoleon who, it seems, was being steadily poisoned with arsenic and mercury most of his life.‡ When

---

*The curve is steepest at 50% so that one can observe there the greatest percentage increase in subjects affected for a small increase in dose. Those who've studied calculus will recognize the dose-effect curve as the integral of a normal probability curve.

†There is also, for most poisons, a *threshold* dose, below which no effect is observed. This will be discussed in Section 5.4.

‡He may have taken them as tonics! See *Neutron Activation Analysis* of this series.

poisons come in slowly enough (when the *dose rate* is low), the body can recover from the damage about as rapidly as it occurs. This seems to be the case with radiation from the natural background. We can apparently recover from 10 rads delivered at the rate of 0.15 rad per year without anything like the damage we might suffer if given 10 rads all at once. Our natural recovery is seemingly treading water quite nicely at 0.15 rad per year.

In much the same way we can usually handle several spaced-out doses much better than a single dose of the same total amount. Some gunslingers of the Old West lived through a half-dozen gunshot wounds in a career, but few survived the full load of a six-gun all at once!

4.42   Then, too, recovery depends on the kind of poison. A man may live through two nearly lethal doses of carbon tetrachloride. This poison is excreted from the body fairly rapidly and, when gone, leaves it without permanent damage. But benzene hydrocarbons can permanently damage the very systems needed for recovery. One severe dose of these may leave the victim anemic for life, greatly reducing his chances for recovery from a second accident. In the same way, the body recovers from low-LET radiation much more rapidly and completely than from high-LET radiation. The latter appears to damage seriously the systems necessary for recovery.

From this one would expect damage from low-LET radiation to increase with increasing dose rate, and this is indeed the case. The rapid administration of a "simple" poison (carbon tetrachloride or low-LET radiation) can "jam" the recovery mechanism badly where slow administration would not. But if the recovery mechanism itself is badly hurt by even small doses of a more complex poison (benzene hydrocarbons or high-LET radiation), an increased dose rate can add little to what has already been done.

4.43   Finally, recovery can be greatly affected by the site of administration. A fireman could live after an intense flame had burnt an arm or a leg badly. The physical isolation of the injured region would permit the uninjured parts of the body to help in the healing of that region. By contrast a burn that was less severe, if spread over the whole body surface, might mean certain death: There would be no uninjured regions to help in the healing. Of course, if the limb were too badly burned its breakdown products could poison the whole body and prevent recovery. Even a localized injury can be too great for any recovery mechanism to handle.

In the same way much more radiation is usually needed to injure an appendage or isolated organ than is needed to cause death if given over the whole body. Still, if the dose to a region of the body is high enough, the effect produced locally can eventually bring about the death of the body.* For example, 10,000 rads to the chest is an LD-50/30 for many mammals, but only 500 rads is an LD-50/30 when delivered over the whole body.

5 CURIOUSER AND CURIOUSER— THE BODILY EFFECTS OF RADIATION

5.1 Worst Things First

5.10 A really intense burst of penetrating radiation (2.5) can actually shock an animal to death. Such a deadly result, however, requires over 100,000 rads† for even the most sensitive mammals. The process produces such wholesale ionization of nerve cell cytoplasm that central-nervous-system function breaks down completely, and the animal dies in convulsions.

5.11 Less penetrating radiation is almost completely absorbed in the first few millimeters of tissue. As a result the effects of a really intense burst are chiefly those of the most violent imaginable sunburn. The animal dies almost as rapidly as above, but the symptoms are those of a burn. There is rapid reddening of the skin (called *erythema*—the animal won't live long enough to blister) and death results from the toxic effects of the burn. Of course, if the radia-

---

*Another example of abscopal effects, see Section 4.2.
†Even if relatively portable radiation sources were used, tons of shielding would be required to protect the operators. Hence the impracticality of "death rays".

tion is confined to a limited area the burn can usually heal itself after blistering, peeling, etc.*

5.12 If penetrating radiation drops into the 10,000-rad range, death is just as certain but comes more slowly. A human exposed to this level of penetrating radiation would be confused and clumsy at first, then lapse into coma and die in a few days. Slightly lower doses usually result in severe lung damage, and the victim dies of a form of pneumonia.

5.13 With less penetrating radiation, effects would be much less severe. The Rongelap natives, who were accidentally exposed in 1954 to severe fallout, received over 2000 rads of beta radiation to their skins.† For the first few days they noticed considerable itching and burning but no erythema. After about 3 weeks dark patches and raised areas appeared, especially on the scalp, along with some loss of hair. Within a few months these healed over com-

---

*I can remember a bromine-82 beta burn of the finger that was more annoying (I couldn't write!) than dangerous, even though the local dose was several thousand rads.

†For more detail about this accident, see *Atoms, Nature, and Man*, a companion booklet in this series.

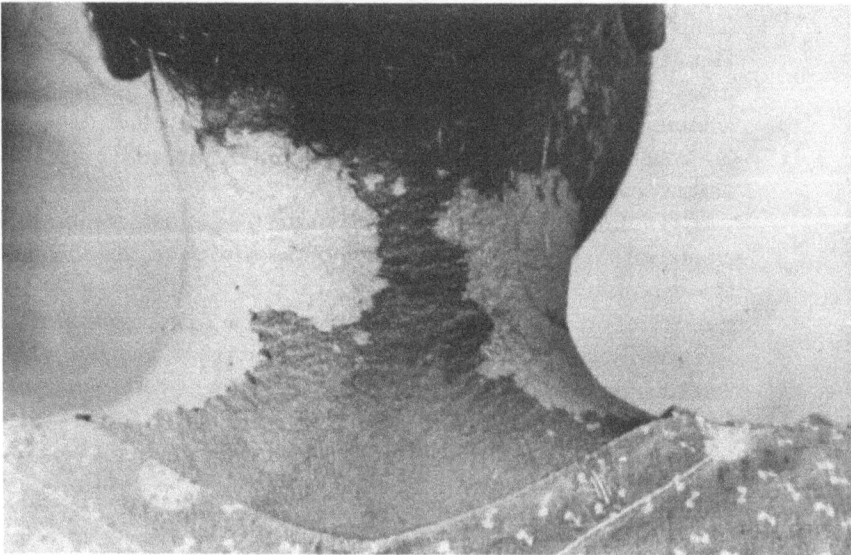

Figure 16 *Changes in pigmentation of skin due to beta burns. On the left is a burn area on the neck of a Rongelap native 1 month after accidental*

pletely as shown in Figure 16. From their experience and from animal studies, we can gather that the human LD-50 for poorly penetrating, external, charged-particle radiation is certainly well over 2000 rads — even if distributed over the entire body surface. Below 2000 rads we find doses of penetrating radiation that leave some hope of survival.

5.2 The LD-50 Region

5.20 The single-dose, whole body LD-50/90 for mammals ranges from about 100 to 1000 rads depending on the type, age, and sex of the mammal and on the type and dose rate of the radiation. Five hundred rads is thought to be an average figure for humans.

5.21 A human exposed in this range develops *acute radiation illness*, and the symptoms of this taken together constitute the *acute radiation syndrome*. Within a dozen hours or so after irradiation, the patient experiences nausea, vomiting, and fatigue, but these pass in a day or two, and the patient feels normal for a few weeks, although many of his blood cells are dying. Eventually the drop in the number of red and white blood cells becomes obvious,

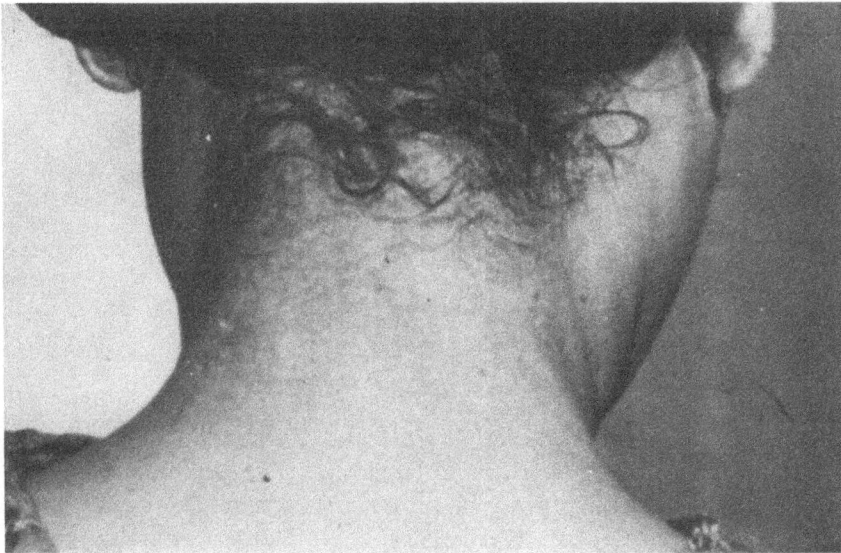

exposure to radiation in 1954. On the right is the same burn area, one year after the accident, showing complete recovery.

along with a drop in the blood platelets. The loss of the red cells (anemia) brings a feeling of weakness because of the poor oxygenation of the blood; its oxygen carriers are dying. The loss of white cells, traditional fighters of infection, leaves the body more open to bacterial invasion. The loss of platelets, which are important to blood clotting, leads to various forms of *hemorrhage* — bleeding from the nose, gums, and even intestines, blood clots in the skin and mucous membranes, and poor wound-healing. If these effects can be overcome in time, the patient will survive. Otherwise he declines steadily until infection, anemia, or hemorrhage kills him.

Figure 17 *These germ-free enclosures are used for studies of radiation effects in small animals. In such an environment, results are not distorted by the animals' acquiring infections caused by germs.*

5.22 The two most sensitive tissues in the body at the LD-50 level are the intestinal wall and the hematopoietic systems (4.16) of the spleen, bone marrow, and lymph nodes. The intestinal wall is important because it is the chief barrier to infection from the bacteria that normally live in the gut, as well as being the primary site of the nutritional absorption needed to sustain the body. If it breaks down, intestinal bacteria can invade the bloodstream directly. If these are not checked by white cells (which would have been destroyed by that amount of radiation) severe and usually lethal *bacteremia* ("blood poisoning") results. Or death can result .from anemia or hemorrhage if the sensitive hematopoietic system is too badly damaged to replace the disappearing blood cells and platelets.

5.23 Fast neutrons and photons differ noticeably from each other in the relative proportion of intestinal or hematopoietic effects in the LD-50 range. Figure 18 shows the survival curves of mice exposed to fast neutrons and to ^{60}Co photons. With fast neutrons most of the animals die around the sixth day after irradiation, and another smaller portion between the tenth and fifteenth days. Examination of the blood and intestines of the animals shows that the six-day deaths were associated chiefly with bacteremia and intestinal-wall breakdown, but that the later deaths were almost exclusively associated with anemia. With the cobalt-60 gamma rays, on the other hand, the 15-day anemia deaths were the predominant ones.

5.24 Associated with the anemia, of course, there is always a temporary destruction of the body's immune response (4.22), since the blood-forming and immune systems are closely linked. This loss of immune response adds insult to the already serious injuries of anemia and in-

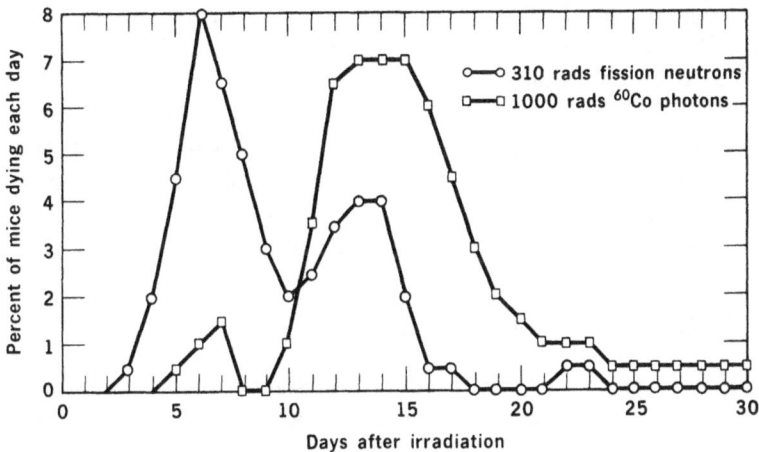

Figure 18 *Studies of mouse survival after single irradiations of fission neutrons or ^{60}Co photons.*

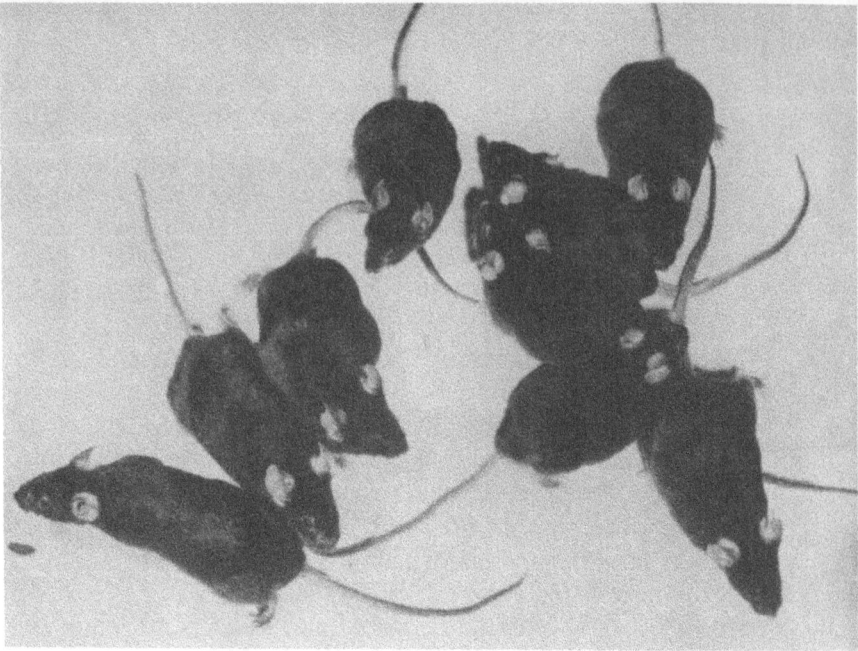

Figure 19 *Two groups of 14-month-old mice that were originally identical. The group on the left was untreated; the group on the right received a large, but not fatal, dose of radiation as young adults. There are only*

testinal breakdown, further lowering resistance to infection.

5.25   In the experiment illustrated by Figure 18, some 650 rads of 250 KVP X rays, 1000 rads of ^{60}Co gamma rays, and 310 rads of fission neutrons* each killed 50% of a group of experimental mice in 30 days. Each of these was an LD-50/30 dose (4.32), in other words. Taking the X rays as a standard, we would say that the ^{60}Co RBE was 650/1000 = 0.65, and the fission neutron RBE was 650/310 = 2.1, for 30-day mouse lethality (2.62). These RBE values are about what might be expected on the basis of LET (2.4), since it is known that neutron LETs are highest, X rays next, and ^{60}Co the least of the three.

5.3 Still Nasty—50 to 400 Rads

5.30   If, because of natural resistance or lowered dose, one survives the acute phase of radiation illness, a number

---

*Fast neutrons produced by the fission of ^{235}U and ranging in energy from about 15 Mev down to nearly zero.

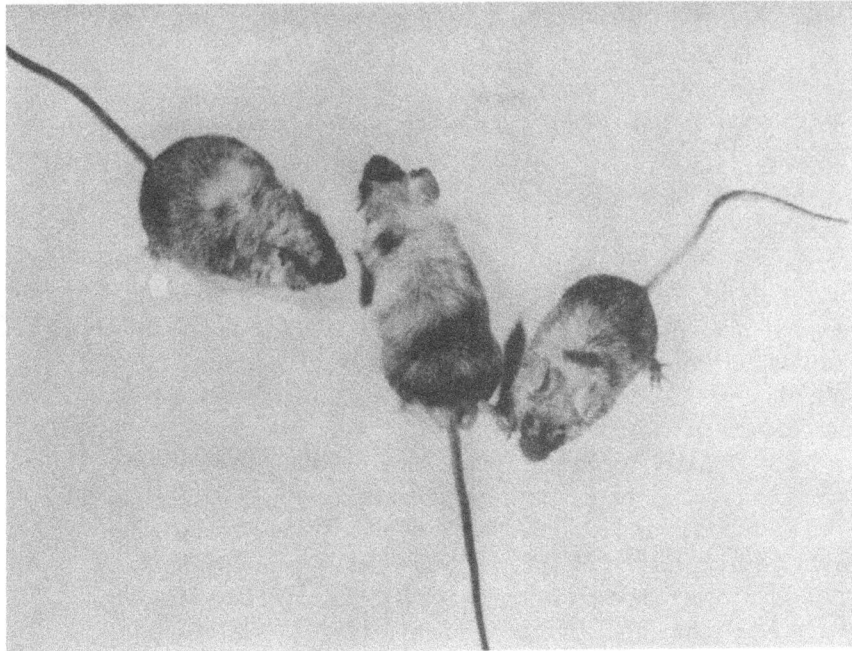

three surviving members of the treated group, and they are gray and senile, while mice in the untreated group are all normal, healthy, and active. Radiation accelerates the aging process and so has been used as an important tool in studies of aging.

of pitfalls still remain. The chief one is the increased probability of cancer, since cells exposed to such disrupting agents as radiation have an increased likelihood of becoming cancerous (4.1).

5.31 *Leukemia,** the earliest and most common cancer associated with radiation begins to show up about a year after irradiation. From various sources† it has been possible to estimate that a single dose of about 200 rads roughly triples one's chances of developing leukemia within a period of 10 years following irradiation. If none shows up in this period a person so exposed is about as unlikely to develop leukemia as anyone else, with a chance of about 0.0005% per year of life.

5.32 When this threat has passed there is still a somewhat increased chance of other forms of cancer, of

*A cancer of the blood, wherein some of the white cells become malignant.
†A-bomb survivors, physicians and scientists who work with radiation, and patients with conditions that required large doses of diagnostic or therapeutic radiation.

later anemia, and *cardiovascular disorders* (heart disease, strokes, etc.) or of developing eye *cataracts* (opaque spots)—perhaps a factor of two for 200-to-400-rad doses. Development of cataracts is much more likely with neutrons than with photons—the neutron RBE being about 10 to 20. If the victim is young enough to be still growing his growth may be impaired.* Fertility is almost always impaired, at least temporarily, and doses of 300 rads to the female ovaries or 1000 rads to the male testes will result in permanent *sterility*.

5.33　Finally a victim may show signs of premature aging—graying hair, skin pigmentation, flabby muscles, "tired blood", and a somewhat lowered disease resistance, with consequent shortened life expectancy (see Figure 19). But, while irradiation is known certainly to "take years off your life", our understanding of this effect is still too poor to allow much prediction of its course or probability.

5.4 Ha, You Missed!

5.40　After radiation at doses below 50 rads, permanent damage of any sort is hard to find. One's chances of developing leukemia or the other late effects just described are probably raised a trifle and one can find some abnormal cells in the body that weren't there before. But, all in all, the potential damage is probably less than what might be expected from smoking, and certainly much less than that from an automobile accident! A single dose of 25 rads is barely enough to cause a noticeable drop in the number of blood *lymphocytes* (one of the white cells), the most radiosensitive cells in the body. And a mouse is a little less prone to run happily around in a mouse wheel (5.8) for a few days after receiving a 25-rad dose. But then a bad cold produces the same temporary effects in a man, and it too lasts only a few days.

5.41　On the other hand, we can never be sure that anything that peppers the cells as indiscriminately as radiation can be completely without effect. Thus, no true *threshold* (a dose below which there is no effect) can be found. But, like

---

*Unlike smoking, radiation will stunt your growth even if you don't inhale.

damage from smoking, poor foods, etc., an effective threshold can and probably does exist — simply in the sense that the damage below that dose is much less than the damage from the 1001 other "natural shocks that flesh is heir to".

5.5 Bits and Pieces

5.50   So far we've concentrated on whole body doses. But, as we noted in 4.2 and 4.4, radiation can be limited to just parts of the body. When this is true, that part can be so sensitive and critical (such as intestine or bone marrow) that its destruction leads to the death of the body at doses

Figure 20 *The Cobalt-60 Gamma Irradiation Room at the Argonne National Laboratory. A single rod of cobalt enriched with ^{60}Co is raised by remote control, permitting irradiation of objects placed about it. Dose rate is controlled by distance from the source.*

around the whole body LD-50. Such parts are called *critical organs* or *critical tissues* and 500 to 1000 rads locally is enough to cause death. An organ can be so sensitive (ovaries) that a few hundred rads ends its usefulness forever. But, if not critical to the overall health of the body, the effects will be limited only to that organ. Then, too, a region can be less sensitive but still critical — and larger doses to it will be needed for lethality. For example, 2000 rads to the head is an LD-50 amount, but it takes 10,000 to the chest to be equally lethal. At the extremes, an arm might tolerate 100,000 rads with no effect worse than damage requiring amputation, or a scalp 20,000 with no effect worse than baldness.

5.51   Finally, a tissue can be quite sensitive, but critical only in the long pull. A thousand rads locally to the lymph nodes would bring about only a temporary radiation illness, but greatly increase one's chances of leukemia in later years. Effects of this sort are usually abscopal (4.21); that is, the final effect is seen elsewhere than at the site of irradiation.

5.52   Now a dose can be received in bits and pieces, not only in space but also in time. What if a body receives a number of small doses or (what amounts to the same thing) is exposed to a low dose rate for a long period of time?

Well, as we noted in 4.4, this sort of stretch-out allows the body time to recover, so that a considerably larger dose is needed to give the same effect. For example, in one animal experiment, 1000 rads of X rays, which is more than needed to kill all the animals in 30 days (140% of the LD-100/30) was given in doses of 250 rads spaced evenly at 2-week intervals. None of the mice died and only about 15% developed leukemia.

5.53   To put it another way, the present stringent rules for protection of workers in radiation laboratories set the *maximum permissible dose* of whole body penetrating radiation at about 250 rads spread over a working career. While such a dose might be felt if given all at once, it is not expected that this same dose, spread out evenly over 30 or 40 years, will have any observable effect.

5.54   The figures just given are for penetrating low-LET radiation, but the same sort of dose rate effect also applies to non-penetrating radiation, too, so long as it is a low-LET type. High-LET radiation, on the other hand, permits much less recovery, as noted in 4.4. Fission neutrons, for example, have been tested on mice at various dose rates ranging downward from 3,000,000 rads per second* to 0.5 rad per week without observable dose-rate effect.

5.55 As a result of this poor recovery rate, the apparent RBE of high-LET radiation rises when late or *chronic radiation illness* effects are considered. Thus 100 rads of fast neutrons will increase the incidence of later leukemia by a factor of 5 or so regardless of how it is delivered. With photons a single dose of 300 rads will have about the same effect but 300 rads delivered at the rate of 0.1 rad a day would have almost no observable effect. In comparing the two we would say that the RBE for neutron-induced leukemia was 3 for single doses but perhaps 20 for chronic (steady or frequent) exposure.

5.56 The observed *chronic effects* of radiation exposure are much the same in either case and are nearly identical to those described in 5.3. But the permissible dose, if these chronic effects are to be avoided, is much lower for high-LET than for low-LET radiation.

5.57 One aspect of chronic low-LET irradiation is worthy of mention, since it's both surprising and, in a sense, amusing. Low levels at low dose rates can have the effect of lengthening life! This is quite commonly observed with laboratory animals, and is sometimes called "the 102% effect", because the life span is increased about 2%. Even more startling is the effect on the flour beetle, *Tribolium confusum*.† His life expectancy is increased about 30% by a dose of 3000 rads delivered at rates up to 10 rads per minute. The explanation probably lies in reduction of infection.‡

---

*These extraordinary dose rates were obtained from bomb explosions, pulsed reactors, and accelerators operating over fractions of a second.

†So-called because he never seems to know where he's going. Of course, in flour, it doesn't matter much.

‡I don't think this means we should all go stand around in the fallout.

5.6 It All Depends

5.60 Up to now we've tacitly assumed a sort of "normal" human body for our doses. In fact, all the figures have been given for healthy, young adults. One would expect the ill and the aged to tolerate radiation less well, and this is, in fact, true. Less obvious is the effect on youth. Since the young are growing, their cells are dividing rapidly. But, as we noted in 4.1, cells are most susceptible to radiation when dividing so that we might expect the young to be more sensitive than adults. This, too, has been confirmed by experiment and by observations of exposed humans. The unborn are most sensitive, as shown in Figures 21 and 22. After birth sensitivity declines as maturity is reached, then rises again as old age is reached. Figure 23 shows this effect for a group of mice.

5.61 Increased body temperature and a higher metabolic rate also seem to increase radiosensitivity over the increase to be expected from increased cell division. The exact nature of these effects, however, is not known very

Figure 21 *A shows a normal chick embryo 10 days after fertilization. C is a 10-day chick that had been irradiated with cobalt-60 gamma rays on the sixth day after fertilization. Note deformities of beak and toes and generalized hemorrhage and swelling. B shows a normal chick embryo 13 days after fertilization. D is a 13-day chick that had been irradiated on the sixth day. In addition to the defects seen in C, there is serious growth retardation.*

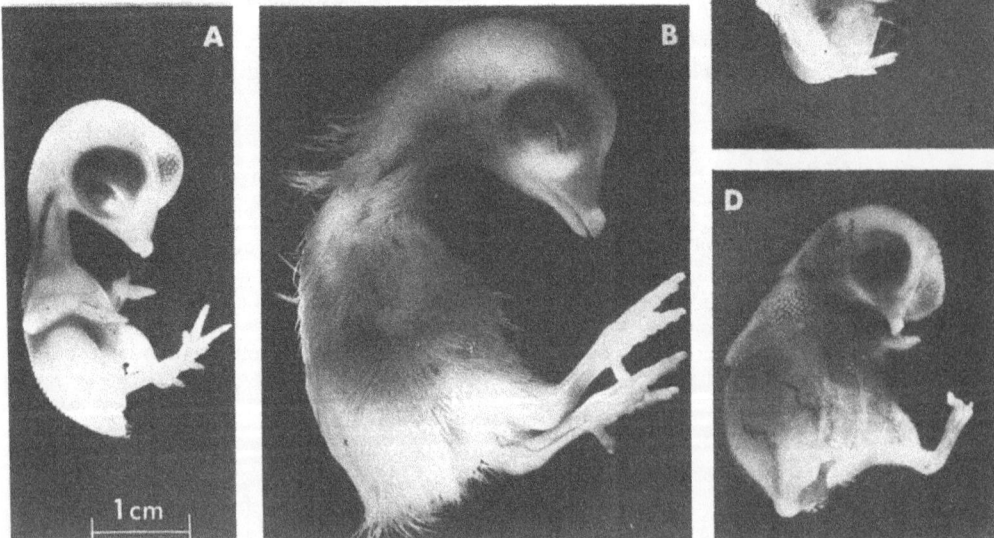

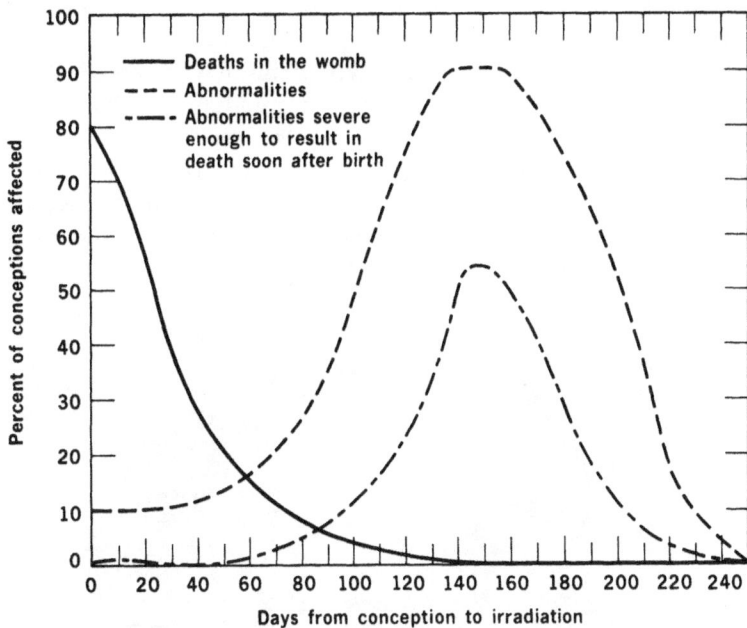

Figure 22 *Probable effects of 250 rads of 250-kilovolt peak X rays during gestation.*

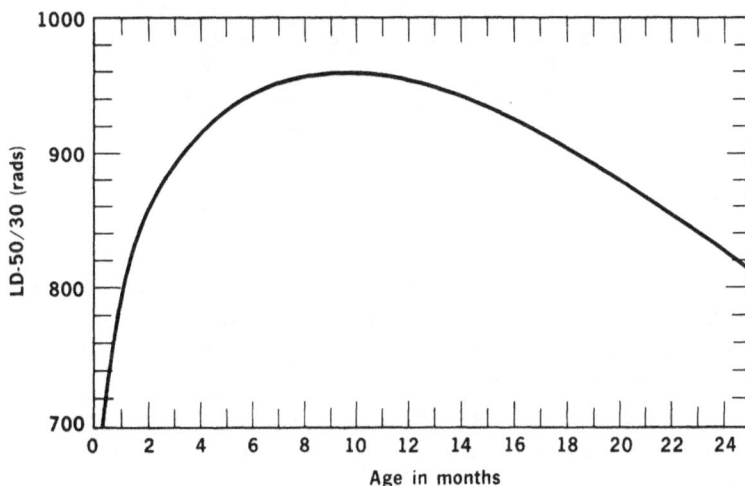

Figure 23 *Sensitivity to cobalt-60 gamma rays as a function of mouse age.*

definitely and certainly is not well understood. One clear effect has been observed, though: *Hibernation* halts radiation damage almost completely. An animal given an LD-50/30, then permitted to hibernate immediately after exposure will die — not within 30 days after exposure but within 30 days after he awakens! This is true even if he hibernates for many months. The reason is probably a combination of lowered body temperature, lowered metabolism, and virtually nonexistent cell division during hibernation.

Figure 24 *A hibernating ground squirrel used in irradiation experiments.*

5.62   Probably the most effective moderator of radiosensitivity is oxygen. Animals irradiated while breathing pure oxygen show a slight increase in sensitivity. But those irradiated while breathing pure nitrogen* show an LD-50/30 nearly 3 times normal for low-LET radiation. For high-LET radiation this *oxygen effect* ranges from small to negligible.

---

*This might seem lethal itself — but even a human can tolerate about 20 minutes of this and still be revived without permanent damage.

5.63   Now this is exactly what might have been expected from the discussion of direct and indirect effects (2.3). High-LET radiation, depending as it does primarily on the direct effect, produces few ions or radicals capable of reacting with oxygen (O_2) to form still more reactive molecules. Thus the presence or absence of oxygen has little effect. But about two-thirds of the damage from low-LET radiation must be of the indirect type, since about this much "vanishes" when oxygen in the tissues disappears. The fact that increasing the oxygen content above that of normal air has little effect shows that normal tissues contain about as much oxygen as the indirect effect can use.

5.64   Finally, as we've mentioned in previous sections, dose effects depend not only on the type of radiation but also on the biological effect being examined (2.4, 2.6, 5.2, 5.3, 5.4). In consequence, RBE values are only roughly constant for a given type of radiation. Just the same, we've listed some representative RBE values, in Figure 25, suggested by the International Commission on Radiation Protection. These should serve as a fair general guide.

Figure 25   RELATIVE BIOLOGICAL EFFECTIVENESS

Radiation type examples	LET, (kev/micron)	Biological effect	Recommended RBE
X, gamma, and beta rays (photons and electrons) of all energies above 50 kev	0.2−3.5	Whole body irradiation, hematopoietic system critical	1
Photons and electrons 10−50 kev	3.5−7.0	Whole body irradiation, hematopoietic system critical	2
Photons and electrons below 10 kev, low-energy neutrons and protons	7−25	Whole body irradiation, outer surface critical	5
Fast neutrons and protons, 0.5-10 Mev	25−75	Whole body irradiation, cataracts critical	10
Natural alpha particles	75−175	Cancer induction	10
Heavy nuclei, fission particles	175−1000	Cataract formation	20

5.7 Fallout Versus Fall In

5.70 As we noted in 3.4 and 5.1, the danger from radioactive nuclides arises chiefly when they become *internal emitters*, that is, when they are absorbed into the body. Even the approximately 2000 rads of external emitter radiation that befell the Rongelap Islanders will probably be of less ultimate importance to them than the much smaller internal doses that they absorbed.

5.71 Since the skin is a relatively poor absorber for most radionuclides, entrance to the body is gained chiefly by inhalation (see Figure 26) and ingestion. Once inside the body the overall effects will depend on the physical and

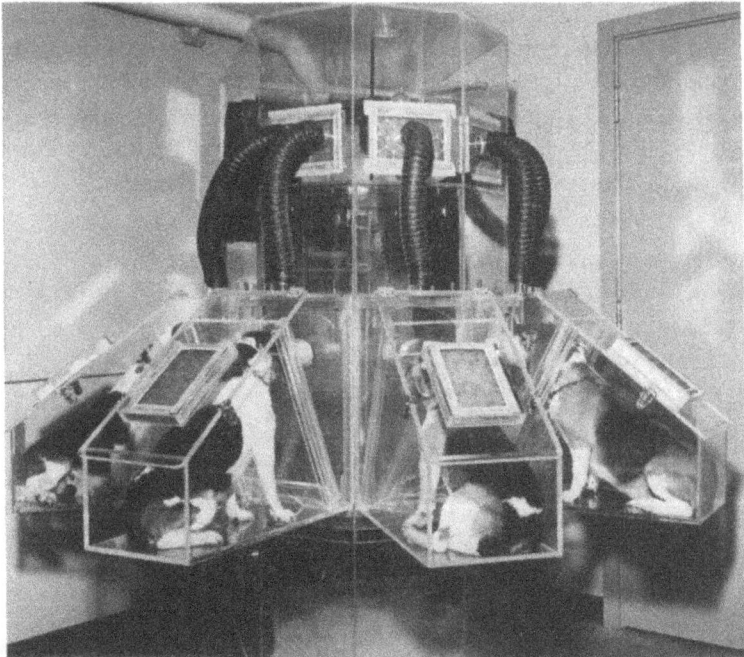

Figure 26 *Studies of the effects of inhaled radionuclides help to determine protective measures to safeguard workers in atomic installations and industry. In the research shown here, the air inhaled by this group of beagles for a measured time contained a small amount of an aerosol containing radioactive plutonium dioxide. Then the animals were examined carefully over a period of up to five years to ascertain whether tissue damage had resulted. The study showed that the body excretes all but the largest particles of the radioactive material.*

chemical properties of the nuclides and on the biochemical and biophysical manner in which it is handled by the body. If the nuclide is in a water-soluble form, it will pass quickly into the blood stream and be distributed throughout the body. If obtained as water-insoluble particles it may largely pass through the gut without absorption or, if inhaled, be trapped in the small spaces of the lungs. Such insoluble particles that do find their way into the body — chiefly via the lung spaces — are slowly scavenged up by the *reticuloendothelial system* (RES) (the lymph system, spleen, liver, etc.) and deposited eventually in the lower gut. In the process they will spend time in the lungs, the RES, and the gut so that each of these will be irradiated to some extent.

5.72   Once a soluble radionuclide is in the blood stream its fate will depend largely on its biochemistry. Sodium, for example, is distributed just about evenly throughout the body. Thus, if radioactive sodium, or another member of its chemical family (potassium, rubidium, cesium, francium), is absorbed it will tend to irradiate the body about as uniformly as external irradiation, and the effects will be much the same. A similar pattern would occur with ^{14}C (as CO_2), ^{3}H (as H_2O), the inert gases (neon, argon, krypton, xenon, and radon). It probably also would be the case with radioactive helium, lithium, nitrogen, or oxygen, except that the latter have no radioisotopes with a half-life longer than a few minutes.

5.73   The rest of the elements do not become so uniformly distributed in the body but, rather, tend to favor certain critical organs (5.5). Like gangs on street corners each tends to look for its own kind. And once "the gang's all here", the local damage is much worse than it would be otherwise. Enough radiation to give 100 ergs to each of the 70,000 grams of a man's body (1 rad) is something quite different from concentrating 700,000 ergs into each gram of a 10-gram lymph node. Although the total ergs to the body are the same, the dose to the lymph node is now 7000 rads. With external radiation this sort of concentration of dose can only be obtained with radiation beams. But with internal emitters it can occur automatically as a result of the normal operation of the body's biochemistry.

5.74   The total damage done will depend on four factors: The LET of the radiation (2.4), the physical half-life (2.8), the quantity of radionuclide absorbed (2.8), and the *biological half-life*. Specific chemical materials are excreted from the body at a relatively constant fraction per unit time (such as 50% per day) so that one can speak of the loss from the body in terms similar to those used for the loss of radioactivity from a radionuclide sample. The biological half-life is the time required for 50% of a given chemical substance to be excreted from the body. Since the two half-lives are controlled by completely independent physical and biological factors, one may find any combination existing for a given radionuclide. For example, cesium-137 and cesium-139 have physical half-lives of 30 years and 9.5 minutes, respectively. But the biological half-life of any cesium in the body is 17 days. Thus, the *effective half-life** of ^{137}Cs would be 17 days but that of ^{139}Cs only 9.5 minutes.

5.75   Having allowed for the effective half-life and for the concentration of radionuclides in critical organs, considerations of quantity and quality still remain. Obviously the greater the number of curies administered (disintegrations per second)† and the greater the energy emitted (ergs per disintegrations), the greater the total energy

---

*Which can be computed from the formula: $\text{Eff} = \dfrac{\text{Phys.} \times \text{Biol.}}{\text{Phys.} + \text{Biol.}}$.

†Of course with an increase in curies comes an increase in dose rate, which, as we saw before, is more important in low-LET than high-LET radiation.

Figure 27   INTERNAL

Nuclide	Half-Lives			Radiations and energies in kev		
	Physical	Biol.	Effective	α	β	γ
^{3}H	12.3 years	19 days	19 days		18	
^{51}Cr	27 days	110 days	22 days			325
^{90}Sr	28 years	11 years	9 years		545	
					2260	
^{226}Ra	1622 years	44 years	44 years	14,500	2700	800
^{239}Pu	24,360 years	120 years	120 years	5,200		
	24,360 years	360 days	360 days	5,200		

(ergs) available for concentration into the mass (grams) of the critical organ. Hence the greater is the total dose (ergs per gram, or rads). Less obviously, but understandably from our previous discussions, the higher the LET of the emitted particles, the higher the RBE, and therefore the greater the biological damage per rad. For this reason one rad to bone from the alpha-emitting (high-LET) bone-seeker ^{226}Ra is much more effective than a rad from the equally bone-seeking but beta-emitting (low-LET) ^{90}Sr. Some of the data concerning internal emitters of interest is summarized in Figure 27. Along with this are the recommendations of the International Commission on Radiological Protection for the *maximum permissible concentrations* (MPCs) that may be allowed in the air or water used by workers in radiation laboratories without expectation of observable injury.

5.76 Once all these factors* have been considered for the injury, the symptoms, and chronic effects of radiation from internal emitters are almost exactly those of the same amount of external radiation delivered to the same critical organs, which we've already discussed. And so we can close the circle of unpleasant radiation effects.

5.8 Spots Before the Eyes

5.80 Can we "feel" radiation? Well, yes and no. Many humans, possibly all, can detect an intense source of

---

*And a few others still too poorly understood to merit mention here.

EMITTERS

Critical organ	Maximum permissible concentrations, curies/cm^3			Form
	Organ	Water	Air	
Whole body	1.5×10^{-7}	2×10^{-7}	10^{-11}	H_2O
Kidney	2×10^{-6}	7×10^{-6}	10^{-11}	Soluble
Bone	10^{-10}	10^{-12}	2×10^{-16}	Soluble
Bone	1.4×10^{-11}	4×10^{-14}	8×10^{-18}	Soluble
Bone	4×10^{-12}	6×10^{-12}	2×10^{-18}	Soluble
Lungs	2×10^{-11}	6×10^{-12}	2×10^{-18}	Insoluble

charged-particle radiation by a tingling of the skin. Both the author and one of his physicist colleagues have accidentally discovered unsuspected "hot" objects in this way. The dose rate required is of the order of 10 rads per hour or so. This sensation is not too useful a detector, especially when so many other things can make you tingle as well or better. The effect is probably one of direct stimulation of the very sensitive touch receptors in the fingertips.

5.81   Of course anyone can see the air ionization caused by an intense beam of charged particles (Figure 1). And fast-moving electrons cause nearly any material to give off light. This *Čerenkov light** is analogous to the sonic boom of a jet plane's shock wave. The electrons, although traveling more slowly than the speed of light in a vacuum, are moving more rapidly than the maximum speed that light would have in the same solid or liquid, hence the

---

*This is the blue glow often seen in color photographs of nuclear reactor cores.

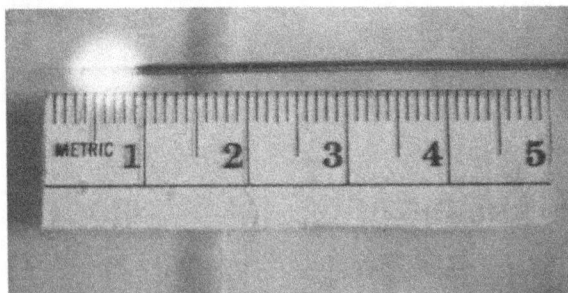

Figure 28   *The bright spot at the tip of the strontium-90 needle below is Čerenkov radiation. This needle has many medical uses, such as destroying pain fibers in the spinal cord to relieve pain (see X-ray photo on right) and destroying the pituitary gland.*

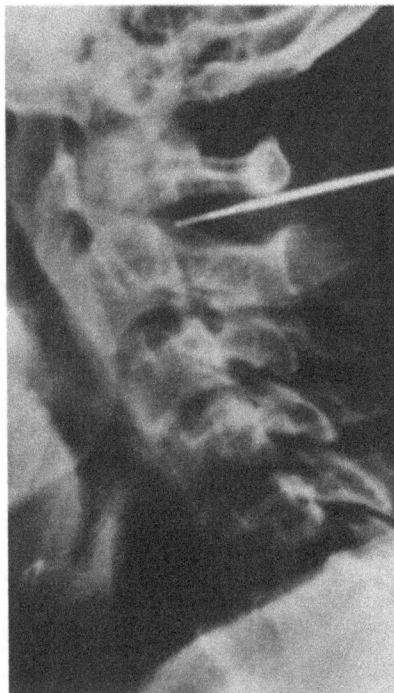

shock-wave effect. After spending 20 or 30 minutes in the dark, most people can see the Čerenkov light given off by even a few millicuries of a beta-emitting radionuclide in a glass bottle. A few substances are so effective in giving off light, for instance the phosphors on watch dials and TV screens, that a few microcuries can be seen in a dimly lit room. But a number of people have reported instances of flashes and bright spots before their eyes during irradiation of the head. Whether this is due to Čerenkov light in the eyeball or to a primary action on the visual cells is unknown.

5.82 It has been shown quite clearly that rats can detect X rays by smell. This ability may be due to a primary effect on the olfactory cells. However, rats also can smell lower concentrations of ozone than humans, and ozone is produced whenever oxygen is ionized by radiation.

5.83 No examples of tasting or hearing radiation are known as yet. But certain hypersensitive mice, which go into convulsions on hearing the ringing of a doorbell, seem to do so much more readily if first exposed to 100 millirads or so of radiation.

5.84 Finally, mice given 25 rads of fission neutrons on a full stomach show a distinct distaste for running around in their little mouse wheels. Sensitivity of the gut to neutron irradiation probably explains this.

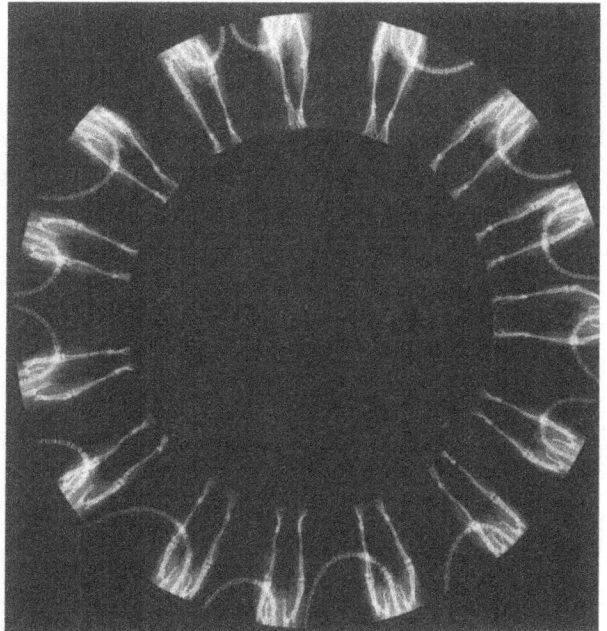

Figure 29 *Fifteen mice in position on an X-ray exposure wheel. Only the hind legs, lower pelvis, and a portion of tail are in the irradiated field.*

5.85 In theory, at least, there is no reason to imagine that radiation cannot stimulate the nervous system, if only by indirect action. One rad will produce about 10^{13} highly reactive ions and radicals in 5 grams of tissue. This is much less than the amount* of "perfume" that a female cockroach need pass on to a 5-gram male to stimulate an active interest on his part!

6 WELL, NOW WHAT?

6.1 An Ounce of Prevention

6.10 As we already mentioned (2.9) the best cure for radiation damage is to avoid it, using either distance or shielding. But, if for any reason avoidance is impractical, what then ?

6.11 Well, from what we've already said one can try to avoid irradiating very young or very old persons, and one can choose forms of radiation with the lowest average doses, the lowest LETs, and the shortest effective half-lives that will still do the job. Also one can shield, or avoid irradiating critical organs even if one cannot protect the whole body. Finally to take advantage of the oxygen effect one can try tolerating a few moments of asphyxiation if need be.

6.12 Of course the latter expedient is usually not a very practical one. But some less drastic alternative ways do exist for achieving the same end. Certain *hormones* and hormone-like chemicals† can reduce the oxygen content of tissues drastically if given in large doses. Others‡ have been found to give specific protection to the white cells, although this may not m e r e l y be an oxygen effect. And nearly any drug that lowers body temperature or metabolic rate (anesthetics, for example) may have a protective ef-fect.§

6.13 Finally, a number of protective compounds have been found, all more or less related to *cysteine,* HS CH_2 CH

---

*The figure reported is 10^{-17} gram, or about 100,000 molecules!

†Such as 5-hydroxytryptamine (serotonin) and para-aminopropio-phenone (PAP).

‡Such as estradiol and similar estrogens.

§Including alcohol!

(NH$_2$) COOH, a normal constituent of body protein. Molecule for molecule cysteine itself is about as effective as any yet discovered. Since the molecules like cysteine contain groups that are known to react well with ions and free radicals, it is believed that the cysteine is sacrificed to the ions and radicals and, by converting these to harmless forms, the process spares the other critical body molecules. Cysteine-like molecules can only reduce the effect of low-LET radiation by about two-thirds and are nearly ineffective against high-LET radiation.

All these *pre-protective agents* are effective only if given before irradiation. Even then the doses needed are so massive that chemical poisoning becomes a serious problem.

6.2 A Pound of Cure

6.20 While little can be done to repair radiation damage directly, much can be done to help along the natural body processes of recovery and repair. Bed rest, good care, and good nutrition are required, of course, for patients suffering from radiation illness. Vomiting and diarrhea can be controlled with drugs, and body-fluid losses made up by transfusions and other means.

6.21 To counteract bacterial invasion of the intestine, antibiotics and disinfectant enemas may be given. Streptomycin has proved particularly useful in this respect and is even more effective if combined with penicillin.

6.22 Hematopoietic damage is something else again, since no simple drug can bring a dead blood-forming cell back to life. Nor would simple infusion of new blood-forming cells from a donor work ordinarily; unless the donor were an identical twin of the recipient, the recipient's immune response to foreign proteins would reject such new cells abruptly. But here radiation deals a curious card, it destroys the normal immune response, at least temporarily (see Figure 30). Bone marrow cells from donors are accepted and will find their way to the victim's hematopoietic system to function there for him. They become part of his immune response "blueprint" and will not be rejected when his immune system begins to function again. A mouse can be irradiated and then given rat blood cells, which seem to function nearly as well as the mouse's own.

6.23 The 8 victims of a 1958 reactor accident at Vinca, Yugoslavia, were all given maximum hospital care and antibiotics and bone-marrow therapy. Despite neutron doses probably well over the LD-50 level all but one of the victims was alive and fairly well years later.

Figure 30 *A patient with acute lymphatic leukemia in a special enclosure that allows safe handling while he is receiving internal radiation from yttrium-90. This procedure is used to suppress the immune mechanism of lymphoid tissue prior to a bone marrow transplant.*

Another note of cheer has been sounded by the discovery that a few hormones can lower the probability of developing eventual cancer in certain tissues, if they are administered soon after irradiation. Others may even help to stimulate recovery of the damaged hematopoietic system.

All in all, maximum care combined with antibiotic and bone marrow therapy can probably raise the LD-50 by a factor of two. It remains to be seen what future research will bring.

6.3 Swords Into Plowshares—Nuclear Medicine*

6.30   Like any powerful tool, radiation can be as much of a blessing as a curse. Medical diagnosis would be set back half a century without the "X-ray picture", whether traditional X rays are used or — the latest wrinkle — photons from radionuclides. Today, too, the ability of tracer radionuclides to follow the chemistry of normal processes is used routinely in the diagnosis and study of dozens of disorders. In the last year well over half-a-million "atomic cocktails" were given for the diagnosis of serious disorders — a tribute to the beneficial use of radiation.

6.31   X rays long provided almost the only alternative to surgery for treatment of cancer and were often chosen for a fair number of other disorders. Today radionuclides are helping this effort along, partly by providing higher energy photons for greater penetration and lower LET at the skin, and partly for their ability to localize in certain tissues. The preferred treatment for the blood disease *polycythemia vera*† involves the administration of ^{32}P, which

---

*Since this will be covered more thoroughly in another booklet of this series, *Radioisotopes in Medicine,* we'll only treat a few highlights here.

†In which an excess of red blood cells results from too many blood-forming cells in the bone marrow.

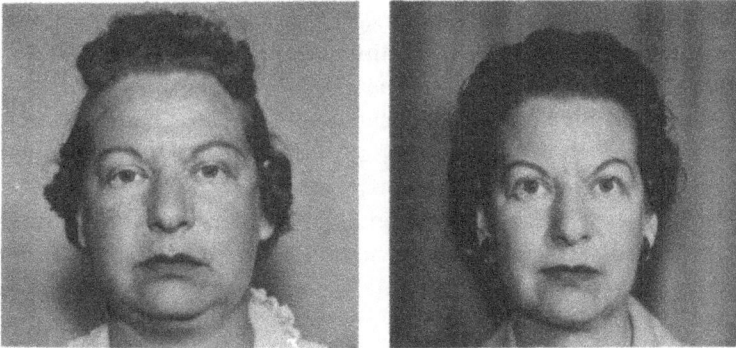

Figure 31 *A patient with Cushing's Disease, a pituitary gland disorder, before treatment (left) and 8 months afterwards (right). She received 8500 rads of 910-Mev alpha particles, delivered to the pituitary over an 11-day period. Five years later she remained apparently free of the disease.*

selectively attacks the offending blood cells. In the same way, radioisotopes of iodine, which localize in the thyroid, are used for the treatment of thyroid cancers.

6.32   The fact that cancer cells usually are rapidly dividing cells and thus more sensitive than other cells, has long been the basis of cancer treatment by radiation. Unfortunately a cancer often grows so rapidly that it leaves its center poorly supplied with blood and oxygen. This core then becomes filled with slowly dividing cells that are not only less radiosensitive but also are protected by the oxygen effect. The problem has been attacked in several ways. One has been the irradiation of the patient while he is breathing pure oxygen at high pressure. This tends to force oxygen into the cancer core, depriving these cells of their oxygen-effect protection but affecting the sensitivity of the other cells only slightly.

6.33   Another approach has combined the lack of oxygen effect with the fact that very high-energy particles start out with low LETs but slow down to higher LETs after penetration of several centimeters of tissue. Very high-energy neutrons, protons, and deuterons have been "beamed" into cancers in such a way that the cancer itself is selectively irradiated with high-LET radiation of high RBE but with little oxygen effect.

6.34   Another treatment utilizes the capture of slow-moving neutrons in otherwise stable nuclides to change the latter into radionuclides. Thus if a fissionable nuclide* can be made to localize within a cancer, low-energy neutrons of a few ev can be beamed into the tumor to cause the release of many Mev of radiation energy. Furthermore, the particles released are of very high LET so that the oxygen effect protection of the core disappears.

6.35   Radioprotective agents have been used to protect parts of the normal body while abnormal parts are being irradiated. But even more intriguing has been the discovery of *radiosensitizing agents* to promote local increases in radiosensitivity in abnormal regions that are to be irradiated.

6.36   In addition many radionuclides have been used in the form of tiny, insoluble spheres, needles (see Figure 28),

---

*To date boron-10 and uranium-235 have been tried, with the latter showing excellent effects in mice.

or particles to bring about the selective irradiation of carefully localized regions without irradiation of the rest of the body. And, finally, it has been suggested, but not yet tried, that neutrons of special energies* could be used to activate (make radioactive) the normally concentrated stable nuclides of certain organs. In this way the stable iodine in the thyroid, for example, could be made radioactive without its floating off into the rest of the body, which is what now happens with the administration of already radioactive iodine.

But, despite all this, the promise of atomic medicine is even greater than its present realization.

---

*Called resonance neutrons because the stable nuclide is "tuned" to capture them selectively.

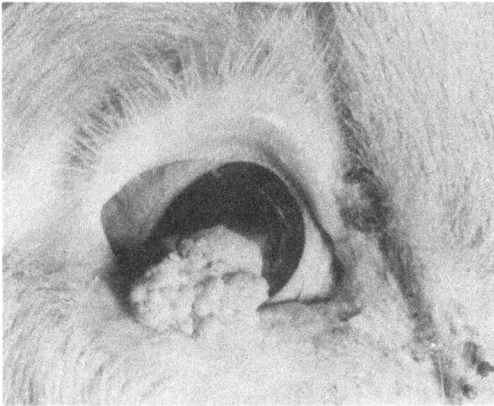

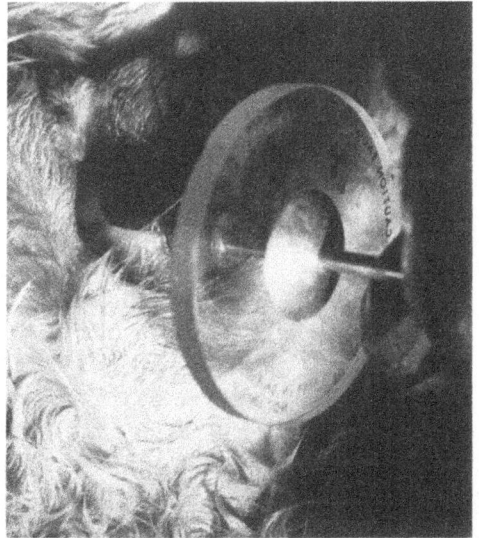

Figure 32 *Use of radioisotopes in veterinary medicine is similar to use for diagnosis and treatment of human disease. Above, an eye cancer in a cow. This type of cancer grows relatively slowly, but if left untreated will extend beyond the point of possible treatment; the animal then must be destroyed. Treatment includes surgical removal of the outer portion, then irradiation with strontium-90, as shown right, above. Because strontium-90 emits beta rays of low penetrating power, the radiation will not damage the lens and deeper structures of the eye. Photo, opposite, shows an eye after successful treatment.*

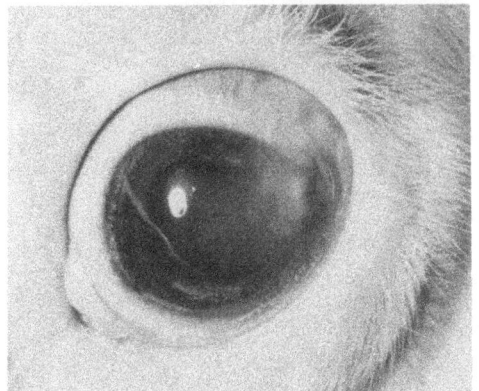

SUMMARY

Our journey through the land of somatic radiobiology will be worthwhile if you carry back a few souvenir ideas. History may prove more than a few of them mistaken, or at least incomplete, but they're the best we have at present.

1. Radiation is a poison probably composed of two parts—direct and indirect action. The latter occurs slowly enough so that some hope exists of blocking it before its damage is complete. Direct action is more like an axe blow—you can avoid it or shield against it but once it has fallen little can be done.

2. Generally speaking, low-LET radiation produces a higher fraction of indirect action than high-LET. High-LET radiation usually brings about more severe and lasting damage.

3. The short-term bodily effects of doses around the LD-50 level are nausea, vomiting, diarrhea, erythema, loss of blood cells, loss of weight, hemorrhage, loss of hair, fatigue, and sterility. The long-term effects are a shorter lifetime, premature aging, increased occurrence of leukemia and other cancers, lowered fertility, cataracts, anemia, and an increased occurrence of cardiovascular disorders.

4. A high enough dose produces certain death; lower doses produce diminishing fractions of short-term and long-term effects. At doses below 50 rads, damage is difficult to ascertain.

5. The damage from low-LET radiation can be lessened by the removal of oxygen, by radio-protective agents, or a lowered dose rate. High-LET radiation is relatively unaffected by such maneuvers.

6. Nearly all bodily radiobiological effects can be understood in terms of critical regions. The DNA of the nucleus is critical to the nucleus, the nucleus is critical to the cell, and certain cells, tissues, and organs are critical to the body.

7. Antibiotic and bone-marrow therapy can reduce markedly the severity of short-term effects. Little has been found to ameliorate long-term effects.

8. Radiation can be as helpful as harmful, and far more lives have been saved by it than lost to it.

GLOSSARY INDEX

A number in italic type following a term listed below indicates the section in which its definition may be found. Usually in that section the term itself will also be set off in italics. Numbers in ordinary type indicate sections in which additional information may be found.

SUGGESTED REFERENCES

Books

Report of the United Nations Scientific Committee on the Effects of Atomic Radiation, General Assembly, 19th Session, Supplement No. 14 (A/5814), United Nations, International Documents Service, Columbia University Press, New York 10027, 1964, 120 pp., $1.50.

The Effects of Nuclear Weapons, Samuel Glasstone (Ed.), U. S. Atomic Energy Commission, 1962, 730 pp., $3.00. Available from the Superintendent of Documents, U. S. Government Printing Office, Washington, D. C. 20402.

The Nature of Radioactive Fallout and Its Effects on Man, Hearings before the Special Subcommittee on Radiation of the Joint Committee on Atomic Energy, Congress of the United States, 85th Congress, 1st Session, U. S. Government Printing Office, 1957, Volume 1, 1008 pp., $3.75; Volume II, 1057 pp., $3.50. Available from the Office of the Joint Committee on Atomic Energy, Congress of the United States, Senate Post Office, Washington, D. C. 20510.

Radiation: What It Is and How It Affects You, Ralph E. Lapp and Jack Schubert, The Viking Press, New York 10022, 1957, 314 pp., $4.50 (hardback); $1.45 (paperback).

Atomic Radiation and Life (revised edition), Peter Alexander, Penguin Books, Inc., Baltimore, Maryland 21211, 1966, 288 pp., $1.65.

Atomic Medicine, Charles F. Behrens and E. Richard King (Eds.), The Williams & Wilkins Company, Baltimore, Maryland 21202, 1964, 766 pp., $18.00.

Atomic Radiation, Radio Corporation of America, 1957, 110 pages, $1.60. Available from RCA Service Company, Government Services, Building 205, Cherry Hill, Camden, New Jersey 08101.

Atomic Energy in Medicine, K. E. Halnan, Philosophical Library, Inc., New York 10016, 1958, 157 pp., $6.00. (Out of print but available through libraries.)

Radiation Biology and Medicine, Walter D. Claus (Ed.), Addison-Wesley Publishing Company, Reading, Massachusetts 01867, 1958, 944 pp., $17.50.

Radiation Injury in Man, Eugene P. Cronkite and Victor P. Bond, Charles C Thomas, Publisher, Springfield, Illinois 62703, 1960, 200 pp., $6.50.

Radiation Protection in Mammals, J. F. Thomson, Reinhold Publishing Corporation, New York 10022, 1962, 220 pp., $8.50.

Atomic Energy Encyclopedia in the Life Sciences, Charles W. Shilling, W. B. Saunders Company, Philadelphia, Pennsylvania 19105, 1964, 474 pp., $10.50.

Articles

Radiation-Imitating Chemicals, Peter Alexander, *Scientific American*, 202: 99 (January 1960).

Radiation and the Human Cell, Theodore T. Puck, *Scientific American*, 202: 142 (April 1960).

Scientific American, 201 (September 1959). This is a special issue on ionizing radiation.

Aerospace Medicine, 36: Section II (February 1965). This is a special issue on radiation biology and space.

Motion Pictures

Available for loan without charge from the AEC Headquarters Film Library, Division of Public Information, U. S. Atomic Energy Commission, Washington, D. C. 20545 and from other AEC film libraries.

The Atom and Biological Science, 12 minutes, black and white, sound, 1953. Produced by Encyclopaedia Britannica Films. This technical film, for intermediate through college audiences, illustrates the effects of radiation on growth and heredity of plants and animals, the use of radioisotopes in tracer studies and in photosynthesis studies, and the methods used to protect the scientists from radiation.

Radiation in Biology: An Introduction, 13½ minutes, black and white or color, sound, 1962. Produced by Coronet Instructional Films. This film, designed for high school science students, explains the meaning of high-energy radiation and shows how this radiation is used in biological research. Radioisotopes are defined and their life is traced from production through use as tools in the study of radiation damage. The effect of radiation on living cells is illustrated by comparisons of irradiated plants and animals with control groups. The effects of radiation on bone marrow, on the protective lining of the intestine, and on chromosomes are also shown.

Radiological Safety, 30 minutes, sound, black and white, 1963. Produced by the Educational Broadcasting Corporation under the direction of the U. S. Atomic Energy Commission's Division of Nuclear Education and Training. This film examines the field of health physics. It defines the maximum permissible limits set by the Federal Radiation Council, the terms, roentgen, rad, and rem, and the RBE (relative biological effectiveness) of radiation. The special laboratory equipment and techniques used to control radiation hazard are described and shown in use.

Medicine, 20 minutes, sound, color, 1957. Produced by the U. S. Information Agency. Four illustrations of the use of radioactive materials in diagnosis and therapy are given: exact preoperative location of brain tumor; scanning and charting of thyroids; cancer therapy research; and the study of blood diseases and hardening of the arteries.

Radiation Protection in Nuclear Medicine, 45 minutes, sound, color, 1962. Produced by the Fordel Films for the Bureau of Medicine and Surgery of the U. S. Navy. This semitechnical film demonstrates the procedures devised for naval hospitals to protect against the gamma radiation emitted from materials used in radiation therapy. However, its principles are applicable in all hospitals.

The following films in the Challenge Series were produced by Ross-McElroy Productions for the National Educational Television and Radio Center under a grant from Argonne National Laboratory. They are each 29 minutes long, have sound, and are in black and white.

Building Blocks of Life (1962) describes free radicals, which are unique fragments of molecules caused by radiation in living systems. These fragments can either kill or seriously damage living cells. The how and why of both the particles and the damage they cause is the topic of this film.

The Immune Response (1962) is concerned with the mechanism by which the body builds antibodies against disease and other foreign substances and with the effects of radiation on this immunizing response.

The following films in the Magic of the Atom Series were produced by the Handel Film Corporation. They are each $12\frac{1}{2}$ minutes long, have sound, and are in black and white.

The Atom and the Doctor (1954) shows three applications of radioisotopes in medicine: testing for leukemia and other blood disorders with radioiron; diagnosis of thyroid conditions with radioiodine; and cancer research and therapy with radiogallium.

The Atom in the Hospital (1961) (available in color and black and white) illustrates the following facilities at the City of Hope Medical Center in Los Angeles: the stationary cobalt source that is used to treat various forms of malignancies; a rotational therapy unit called the "cesium ring", which revolves around the patient and focuses its beam on the diseased area; and the total-body irradiation chamber for studying the effects of radiation on living things. Research with these facilities is explained.

Atomic Biology for Medicine (1956) explains experiments performed to discover effects of radiation on mammals.

Atoms for Health (1956) outlines two methods of diagnosis and treatment possible with radiation: a diagnostic test of the liver, and cancer therapy with a radioactive cobalt device. Case histories are presented step-by-step.

Radiation: Silent Servant of Mankind (1956) depicts four uses of controlled radiation that can benefit mankind: bombardment of plants from a radioactive cobalt source to induce genetic changes for study and crop improvement; irradiation of deep-seated tumors with a beam from a particle accelerator; therapy of thyroid cancer with radioactive iodine; and possibilities for treating brain tumors.

www.ingramcontent.com/pod-product-compliance
Lightning Source LLC
Chambersburg PA
CBHW032015190326
41520CB00007B/478

* 9 7 8 1 4 7 9 4 2 5 7 3 0 *